Satellite image of southern Perú and adjacent Bolivia. The Río Madre de Dios is the large river running from the left toward the upper right corner. The Río Heath runs from the bottom toward the top just right of center. It separates the Peruvian pampas from the more extensive Bolivian pampas (pink/purple in this image). The Río Tambopata runs up from the left side and joins the Río Madre de Dios at Puerto Maldonado. Image from EOSAT, LANDSAT Thematic Mapper, bands 3, 4, and 5. July 27, 1991.

Rapid Assessment Program

RAP

Working

Papers

6

The Tambopata-Candamo Reserved Zone of Southeastern Perú: A Biological Assessment

Robin B. Foster
Theodore A. Parker III
Alwyn H. Gentry
Louise H. Emmons
Avecita Chicchón
Tom Schulenberg
Lily Rodríguez
Gerardo Lamas
Hernan Ortega
Javier Icochea
Walter Wust
Mónica Romo
J. Alban Castillo
Oliver Phillips
Carlos Reynel
Andrew Kratter
Paul K. Donahue
Linda J. Barkley

CONSERVATION INTERNATIONAL

NOVEMBER 1994

RAP Working Papers are published by:

Conservation International

Department of Conservation Biology

1015 18th Street, NW, Suite 1000

Washington, DC 20036

USA

202-429-5660

202-887-0193 (fax)

Managing Editor: John L. Carr

Assistant Editor: Kim Awbrey

Design: KINETIK Communication Graphics, Inc.

Maps: Chris Rodstrom, Andrew Waxman, Ali Lankerani

Cover photograph: Frans Lanting

 Printed on recycled paper in the
United States of America

CONSERVATION INTERNATIONAL   **Rapid Assessment Program**

Table of Contents

Participants

RESEARCH TEAM

Theodore A. Parker III
Ornithologist, RAP Team Leader
Conservation International

Alwyn H. Gentry
Botanist
Conservation International

Robin Foster
Plant Ecologist
Conservation International

Louise H. Emmons
Mammalogist
Conservation International

Gerardo Lamas
Lepidopterist
MHN-UMSM

Hernan Ortega
Ichthyologist
MHN-UMSM

Lily Rodríguez
Herpetologist
MHN-UMSM

Mónica Romo
Mammalogist
University of Missouri at St. Louis

César Ascorra
Mammalogist
MHN-UMSM

Carlos Reynel
Botanist
Missouri Botanical Garden

J. Alban Castillo
Botanist
MHN-UMSM

Javier Icochea
Herpetologist
MHN-UMSM

Walter Wust
Ornithologist
ACSS

Percy Nuñez
Botanist
ACSS

Rosa Ortiz
Botanist
Missouri Botanical Garden

Andrew Kratter
Ornithologist
Louisiana State University

TRIP COORDINATORS

Carlos Ponce
Vice President for Andean
Region Programs
Conservation International

Liliana Campos
Director, Andean Region Programs
Conservation International

Avecita Chicchón
Anthropologist,
Director, CI-Peru
Conservation International

Brent Bailey
Rapid Assessment Program Coordinator
Conservation International

OTHER CONTRIBUTORS

Thomas S. Schulenberg
Ornithologist
FMNH

Paul K. Donahue
Ornithologist

Oliver Phillips
Botanist
Missouri Botanical Garden

Linda J. Barkley
Mammalogist
Louisiana State University

EDITORS

Robin B. Foster

John L. Carr

Adrian B. Forsyth

Organizational Profiles

CONSERVATION INTERNATIONAL

Conservation International (CI) is an international, nonprofit organization based in Washington, D.C., whose mission is to conserve biological diversity and the ecological processes that support life on earth. CI employs a strategy of "ecosystem conservation" that seeks to integrate biological conservation with economic development for local populations. CI's activities focus on developing scientific understanding, practicing ecosystem management, stimulating conservation-based development, and assisting with policy design.

Conservation International
1015 18th St. NW, Suite 1000
Washington, DC 20036 USA
202-429-5660
202-887-5188 (fax)

CI - Perú
Calle Chinchón 858-A
San Isidro
Lima 27
PERÚ
5114-408-967 (phone/fax)

MISSOURI BOTANICAL GARDEN

The Missouri Botanical Garden, a nonprofit organization founded in 1859, is the oldest continually operating botanical garden in the United States. Its mission is to discover and share knowledge about plants and their environment, in order to preserve and enrich life. The major projects of the Research Division, with a staff of over 150, center on studies of diversity, classification, conservation, and uses of plants, especially in the tropics. The Garden houses a rapidly expanding herbarium, currently containing more than 4 million collections from around the world, and a library with approximately 115,000 volumes.

Missouri Botanical Garden
P.O. Box 299
St. Louis, MO 63166-0299 USA
314-577-5100
314-577-9596 (fax)

MUSEO DE HISTORIA NATURAL UNIVERSIDAD NACIONAL MAYOR DE SAN MARCOS

The Natural History Museum is a branch of the Biology Department of the University of San Marcos in Lima, a government institution founded in 1918. Since its creation, the museum has contributed greatly to the scientific knowledge of the fauna, flora and geology of Peru. The museum's main goal is the development of scientific collections, systematic research, and to provide the data, expertise and human resources necessary for understanding Peru's biogeography and promoting conservation of Peru's many ecosystems.

The museum conducts field studies in the areas of Botany, Zoology, Ecology, and Geology-Paleontology. Each of the fifteen departments has its own curator, associated researchers and students. Over the past 10 years the museum has conducted intensive fieldwork in different protected areas in lowland forest such as Manu National Park, Abiseo National Park, and Pacaya-Samiria National Reserve.

Museo de Historia Natural
Universidad Nacional Mayor de San Marcos
Apartado 14-0434
Lima - 14
PERÚ
51-14-710117 (phone)
postmaster@musm.edu.pe (e-mail)

Acknowledgments

We would like to thank the National Parks Division of the Ministry of Agriculture, and the Directors and staff of the Pampas del Heath National Sanctuary and the Tambopata Candamo Reserved Zone. Special thanks to Ing. Jorge Ugaz (then Chief of National Parks Division), Ing. Manuel Uceda (then Head of the Dirección de Flora y Fauna), Rosario Acero (granted permits for collections), Ricardo Gutierrez (helped acquire permits), Fernando Rubio (Director of the Pampas del Heath National Sanctuary and FPCN) and Oscar Rada (Director of the Tambopata-Candamo Reserved Zone).

Fieldwork for this report would not have been possible without the help of the field staff of the Pampas del Heath National Sanctuary, especially Dario "Taco" Cruz, Vicente Vilca, Juan Racua, Armando Cruz, Jose Luis de La Cruz, and the park guards of the sanctuary. For his help at the Ccolpa and Cerros del Távara we would like also to extend thanks to Eduardo Nycander.

The Municipality of Puerto Maldonado, CANDELA, and FENAMAD all provided assistance with equipment that helped make the trip a sucess.

Many special thanks to Max Gunther and Explorer's Inn for their help and hospitality in the field. Marcia Morrow, Orlando James, Enrique Ortiz and the resident naturalists there were extremely supportive and helpful in this project.

We are grateful to the Ministry of Agriculture's Programa de Parques Nacionales and CI's Carlos Ponce for their organization of the Tambopata workshop, and for their guidance and support of this project. Bettina Torres provided important assistance in the early planning stages.

We thank the Museo de Historia Natural of the Universidad Nacional Mayor de San Marcos (UNMSM) for their collaboration, especially Anselmo Turpo, José Luis de Coloma, Ignacio de la Riva, and Lily Rodriguez. Robert P. Reynolds (USNM) identified the amphibian and reptile material deposited at the US National Museum of Natural History; Charles Meyers (AMNH) collaborated in confirming the new species of *Epipedobates* and J. Icochea (MUSM) shared his field notes from the TREE expedition with Lily Rodríguez. At the Smithsonian we thank Drs. M. Carleton and A. Gardner of the National Museum of Natural History and U.S. Fish and Wildlife Service for helping to identify mammal specimens. Thanks to the Missouri Botanical Garden and to the Field Museum of Natural History Botany Department staff for thier assistance with collections.

Very special thanks to Jaqueline Goerck and Tom Schulenberg for their help in the compilation of Ted Parker's data and for their neverending support and advice on this report. Rosa Ortiz was instrumental in providing the material written by Al Gentry.

At the CI office in Washington, we thank Liliana Campos and Brent Bailey for coordination of the trips; Ali Lankerani, Chris Rodstrom and Andrew Waxman for mapping, Kim Awbrey and Adrian Forsyth helped in assembling the report, and Enrique Ortiz for comments and review of the report.

Frans Lanting, Andrew Kratter, Louise Emmons, and Kim Awbrey graciously provided some of the photographs for the report.

We are most grateful to the John D. and Catherine T. MacArthur Foundation for providing the financial support for the expedition and publication of this report.

Foreword

The waters of the Tambopata River begin with the melting snows in the high Andes of Southern Perú. The waters flow north towards the Amazonian lowlands in a rapid descent through saw-tooth ridges. The cloud forests of this region are dense with every limb matted with fern, orchid and moss and the only trails are those of the secretive spectacled bear and illusive mountain lion. The forest grows taller as the river runs through the foothills and then begins its shifting meander into the lowlands.

Along the banks of the Tambopata brilliant flocks of macaws concentrate by the hundreds to feed on mineral-rich soil pockets. Giant otters hunt the rivers for enormous catfish. Vast expanses of forest extend in all directions.

This great swath of forest and river running from the "ceja de montaña" down to Amazonia is among the most diverse and important expanses of tropical wilderness. This landscape is a vital reservoir of Peru's biological riches and, because Peru itself is home to such a high percentage of life on earth, the Tambopata-Candamo region is of global conservation significance.

The Tambopata-Candamo region is strategically located in the heart of a potential conservation "mega-corridor" that would biologically link conservation units in Perú with those of Bolivia.

This corridor and large units such as Tambopata-Candamo offer the best hope for retaining viable populations of animals such as giant otters and other species with large space requirements. Tambopata-Candamo is also now well documented as having a highly diverse flora and fauna with many endemic species.

This publication describes the biological importance of Tambopata-Candamo and makes recommendations regarding the conservation and development of the region. The investigations lead to a clear conclusion. A protected area in Tambopata-Candamo can serve as a key element in the sustainable development of the region and as a priceless resource for the future of Perú.

Two of the authors of this report, two people who knew Tambopata-Candamo well—the legendary ornithologist Ted Parker and the prodigious botanist Al Gentry—are now gone. They died August 3, 1993 while working towards their life-long dream: the understanding and conservation of tropical diversity. Tambopata-Candamo was a large part of their dream and it is tragic they will not see the creation of Tambopata-Candamo as a national park. But this dream remains very much still alive. We hope this report will further the goal of establishing Tambopata-Candamo as a protected area for the people of Perú and the world, for the memory of Ted and Al, for all time.

Adrian Forsyth
Director of Conservation Biology

Overview

INTRODUCTION

In May-June 1992, biologists from Conservation International's Rapid Assessment Program (RAP) team and counterparts from Peruvian institutions undertook an evaluation of the principal ecosystems along the upper Río Tambopata and Río Távara, and the Pampas del Heath. The primary goals of this effort were to provide a clearer picture of the full range of biological diversity that occurs in this region, and to apply that knowledge to the zoning process. The group inventoried plants, mammals, birds, reptiles and amphibians in three areas: (1) on the floodplains and adjacent uplands near the Ccolpa de Guacamayos, on the Río Tambopata about halfway between the mouth of the Río Távara and the mouth of the Río Malinowski; (2) on several low Andean foothill ridges (ca. 500-900 m in elev.) that rise above the Río Távara between the point where it meets the Río Tambopata and its confluence with two smaller rivers higher upstream (i.e., the Ríos Guacamayo and Candamo); and (3) the Pampas del Heath, where butterflies and fish were included in the inventory efforts. We were anxious to obtain baseline data on the structure and composition of plant and vertebrate communities in these areas in order to understand how these communities vary, both in the lowlands along the Río Tambopata and also along elevational gradients in the Andean foothills. From the outset, we planned to compare our findings with those of the many scientists (including ourselves) who have been working in the Explorer's Inn Reserve

INTRODUCCIÓN

Durante Mayo y Junio de 1992, un equipo de biólogos del Programa de Evaluación Rápida (RAP siglas en inglés para Rapid Assessment Program) de Conservation International y la contraparte de instituciones peruanas llevaron a cabo una evaluación de los principales ecosistemas a lo largo del alto Río Tambopata y Río Távara, y en las Pampas del Heath. Las metas iniciales de este esfuerzo fueron las de proveer una visión mas clara de la diversidad biológica que ocurre en esta región, y aplicar estos conocimientos para el proceso de zonificación. El grupo inventarió plantas, mamíferos, aves, reptiles y anfibios en tres áreas: (1) en los llanos inundados y tierras altas adyacentes cerca a la Ccolpa de Guacamayos, en el Río Tambopata que se encuentra entre la boca del Río Távara y la boca del Río Malinowski; (2) en varias zonas al pie de la Cordillera de los Andes (500-900 m. de elevación) que se levantan sobre el Río Távara entre el punto donde éste se encuentra con el Río Tambopata y su confluencia con dos pequeños ríos "río arriba" (los Ríos Guacamayo y Candamo); y (3) las Pampas del Heath, donde mariposas y peces fueron incluídos en los inventarios. Estuvimos ansiosos de obtener datos iniciales de la estructura y composición de las comunidades de plantas y vertebrados en estas áreas para poder entender cómo estas comunidades varían, tanto en las zonas bajas a lo largo del Río Tambopata y como también a lo largo de las gradientes altitudinales al pie de los Andes. Desde el principio planeabamos comparar

(Tambopata Reserved Zone), along the lower Río Tambopata, between 1974 and the present. The latter area is one of the best known research sites in upper Amazonia. Forests within approximately 5 km of the mouth of the Río La Torre are among the most species-rich communities yet reported on Earth for several taxonomic groups, including birds and dragonflies.

The Current Situation in the Tambopata-Candamo Reserved Zone

Southeastern Perú and adjacent parts of Bolivia contain what is probably the largest and least disturbed area remaining of Upper Amazonian and Lower Andean ecosystems. In Perú, the Tambopata-Candamo Reserved Zone (TCRZ) was created in January 1990 (Figs. 1 and 2). According to Peruvian law, a reserved zone is a transitory protected area that allows for planning to determine other, permanent land use categories. Since 1990, Conservation International has collaborated with the Peruvian government to carry out ecological and social studies in order to assess current and potential land uses in the region. As part of this integrated effort, the RAP team visited the Tambopata and Heath regions in 1992. All the information gathered in this process has been analyzed by local citizen groups, government officials, and nongovernmental organization representatives in workshops, fora and informal meetings. This participatory planning process resulted in the proposal of a strict protected area: the Tambopata-Heath National Park, or **Bahuaja-Sonene** in the Ese'eja language (INRENA 1994). The Peruvian Natural Protected Areas office has proposed to integrating the Pampas del Heath National Sanctuary with the Tambopata core protected area into a 968,587.25 ha national park. At present, this proposal is being considered by higher au-

nuestros datos con aquellos de los muchos científicos (incluyéndonos nosotros mismos) quienes, desde 1974 al presente, hemos estado trabajando en la Reserva Explorer's Inn (Zona Reservada de Tambopata), a lo largo del bajo Río Tambopata. Esta reserva es una de las áreas más investigadas de la Amazonía alta (cercana a los Andes). Los bosques aproximadamente a 5 Km de la boca del Río La Torre presentan las comunidades mas ricas en especies para algunos grupos taxonómicos (incluyendo aves y libélulas), al presente reportadas en la Tierra.

La Situación Actual en la Zona Reservada Tambopata-Candamo

Al sudeste del Perú y partes adyacentes de Bolivia quedan áreas que probablemente son los más grandes y menos perturbadas, de los ecosistemas de la Alta Amazonía y parte baja de los Andes. En el Perú, la Zona Reservada Tambopata-Candamo (ZRTC) fue creada en Enero de 1990 (Figs. 1 y 2). De acuerdo a las leyes peruanas, una zona reservada es un área protegida transitoriamente que permite planear y determinar otras categorías de uso de tierras permanente. Desde 1990, Conservacion Internacional (CI) ha colaborado con el gobierno peruano en el estudio ecológico y social para evaluar el uso actual y potencial de la tierra en la región. Como parte de este esfuerzo integrado, el equipo del RAP visitó las regiones de Tambopata y del Heath en 1992. Toda la información obtenida en este proceso ha sido analizada por grupos de ciudadanos locales, oficiales del gobierno, y representantes de organizaciones no-gubernamentales en talleres, y en reuniones formales e informales. Este proceso de planeamiento con amplia participación ha dado como resultado que se proponga un área estrictamente protegida: El Parque Nacional Tambopata-Heath, o

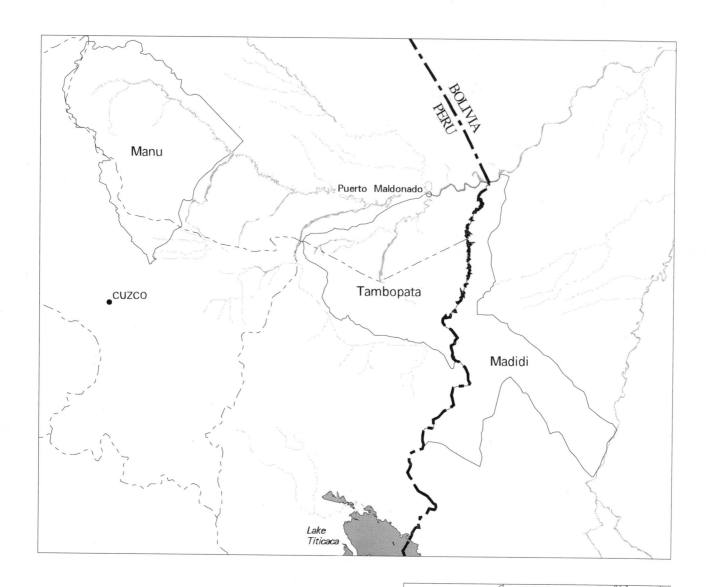

Figure 1. *Map of Perú showing the Manu Biosphere Reserve, the Tambopata-Candamo Reserved Zone, and the proposed Alto Madidi National Park in Bolivia*

CONSERVATION INTERNATIONAL

Rapid Assessment Program

thorities within the Peruvian government.

Currently, no people live within the borders of the proposed park. There are, however, indirect uses of the resources within this area. Ecotourism takes place, and some hunting and fishing is practiced by indigenous groups. Recent surveys suggest that about 3,200 people live in the buffer area in the TCRZ-Madre de Dios (north of the core area), while there are about 3,800 people in the TCRZ-Puno (south of the core area). The ethnic composition of these two groups and their subsistence practices differ. Most of those living in Madre de Dios are second or third generation migrants from other areas of Perú (mainly the Andes). They have learned subsistence practices from neighboring indigenous groups. These farmers form part of the Agrarian Federation of the Department of Madre de Dios (FADEMAD) that have learned from past mistakes in resource use and are now putting into practice concepts of sustainable agriculture and agroforestry through conservation-based development projects. These farmers were initially critical of conservation issues and the validity of protected areas. Today, due to their participation in the planning process for Tambopata, they have realized that renewable resources may be finite if not used wisely and they have become active partners in conservation planning for the region.

The Ese'eja people are a native Amazonian group from the Tacana linguistic family. Their traditional territory included the Tambopata and Madidi regions between the Peruvian and Bolivian borders. Some researchers estimate that the Ese'eja numbered nearly 10,000 people at the beginning of this century. However, because of western diseases, migration and the impact of rubber exploitation, there are now only 600 Ese'eja left within Peruvian territory. They live in three communities inside Perú (there are oth-

Bahuaja-Sonene en el lenguaje Ese'eja (INRENA 1994).

La Oficina peruana de Areas Naturales Protegidas ha propuesto integrar el Santuario Nacional de las Pampas del Heath con el núcleo del área protegida de Tambopata dentro de los 968,587.25 hectáreas del Parque Nacional. Al presente, esta propuesta esta siendo considerada por las autoridades más altas dentro del gobierno peruano.

Actualmente, no hay poblaciones humanas que vivan dentro de los límites del propuesto parque. Sin embargo, si hay un uso indirecto de los recursos dentro de esta área. Algo de caza y pesca es practicado por grupos indígenas, y tambien se lleva a cabo Ecoturismo. Encuestas recientes sugieren que cerca de 3,200 personas viven en el área de amortiguamiento en la ZRTC-Madre de Dios (al norte del área nucleo), mientras que cerca de 3,800 personas viven en la ZRTC-Puno (al sur del área nucleo). La composición étnica de estos dos grupos y sus prácticas de subsistencia difieren. La mayoría de los que viven en Madre de Dios son segunda y tercera generación de migrantes de otras areas del Perú (principalmente de los Andes). Ellos han aprendido las prácticas de subsistencia de los grupos indígenas vecinos, ademas de la experiencia propia de varios años. Estos agricultores forman parte de la Federación Agraria del Departamento de Madre de Dios (FADEMAD) que han aprendido de errores del pasado en el uso de recursos y que hoy ponen en práctica conceptos del agricultura sostenida y agroforestería a través de proyectos de desarrollo basados en la conservación.

Estos agricultores fueron inicialmente críticos de los temas de conservación y del valor de las áreas protegidas. Hoy en día, en parte debido a su participación en el proceso de planeamiento para Tambopata, ellos se han dado cuenta que los recursos renovables

Conservation-

ists must work

closely with

local people

in order to

find viable

economic,

social, and

biological

alternatives ....

ers in Bolivia): one on the Río Tambopata (Infierno) and two in the Heath region (Palma Real and Sonene). The Federation of Indigenous Peoples from Madre de Dios (FENAMAD) proposed that the park name should be the Ese'eja names of the two rivers that sustained the people that originally inhabited this high biodiversity area (Bahuaja = Tambopata, Sonene = Heath).

The southern portion of the Reserved Zone, the upper Tambopata river region, is inhabited seasonally by Aymara and Quechua peoples that migrate to this region to cultivate coffee. The other part of the year these people live around Lake Titicaca at 4,000 meters altitude. Although the region around the town San Juan del Oro has been occupied since pre-Hispanic times for the exploitation of gold, early in this century the region attracted more people for the extraction of rubber (*Castilla* sp.), quinine (*Cinchona* spp.) and other forest products. More recently, coffee cultivation has been the most important cash generating activity for farmers in the region; however, due to the low international price of coffee coupled with high transportation prices and political turmoil, this economic activity has not yet yielded much revenue for producers in the region. The understanding of all these socioeconomic variables has led us to tailor conservation planning workshops in Puno in a different way than those in Madre de Dios. Local peoples' responses have been favorable so far because their own experiences have taught them that they have to rely primarily on their own capabilities and not on external subsidies for answers to their resource use problems.

The conservation planning process in Tambopata is not yet finished. It is not enough to have a national park declared. Conservationists must work closely with local people in order to find viable economic, social, and biological alternatives that allow for the pro-

pueden ser limitados si no son usados prudentemente y ahora son socios activos en la planificación de la conservación de la región.

La población Ese'eja es un grupo nativo Amazónico de la familia lingüística Tacana. Su territorio tradicional incluye las regiones de Tambopata y Madidi entre las fronteras peruana y boliviana. Algunos investigadores estiman que la población Ese'eja al comienzo de siglo fue de cerca de 10,000. Sin embargo, debido a enfermedades del mundo occidental, migración y al impacto de la explotación del caucho, ahora quedan solamente 600 Ese'eja dentro del territorio peruano. Ellos viven en tres comunidades dentro del Perú (hay otras en Bolivia): Una en el Río Tambopata (Infierno) y dos en la región del Heath (Palma Real y Sonene). Fue la Federación de Poblaciones Indígenas de Madre de Dios (FENAMAD) la que sugirió que el nombre del Parque debería llevar el nombre en Ese'eja de los dos ríos que sostuvieron a las poblaciones que originalmente habitaron estas áreas de gran biodiversidad (Bahuaja= Tambopata, Sonene=Heath).

La porción sur de la Zona Reservada, la región alta del río Tambopata, es habitada temporalmente por poblaciones Aymara y Quechua que migran a esta región para cultivar café. Durante la otra parte del año estas poblaciones viven alrededor del Lago Titicaca a 4,000 metros de altitud. Aunque la región alrededor del pueblo San Juan del Oro ha sido ocupada desde tiempos pre-Hispánicos por la explotación del oro, a principios de este siglo la región atrajo más gente por la extracción del caucho (*Castilla* spp.), quinina (*Cinchona* spp.) y otros productos del bosque.

Mas reciente, es el cultivo de café lo que ha venido a ser la actividad más importante generando ingresos para los agricultores en la región; sin embargo debido al bajo precio internacional del café unido al alto costo de transporte y además a la inestabilidad política,

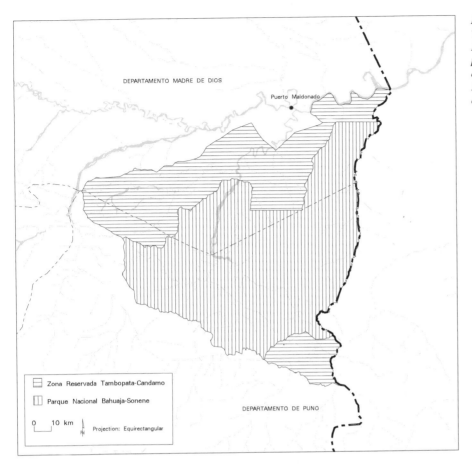

Figure 2. Map of the Tambopata-Candamo Reserved Zone with the proposed boundaries of Bahuaja-Sonene National Park (based on INRENA 1994).

Los conserva-cionistas deben trabajar muy cerca con la población local para poder encontrar alternativas viables tanto económicas, sociales y biológicas ....

tection of biodiversity and the sustainable use of natural resources. This is the challenge we must meet if we are to retain the biological wealth of Tambopata-Candamo.

SUMMARY OF RESULTS

We focused on two main areas of conservation importance that are contiguous at the base of the Andes in southeastern Perú, and adjacent to the least disturbed forest in Bolivia, which has been proposed for inclusion in Madidi National Park:

1 The Río Tambopata and the Cerros del Távara (Távara Mountains). Although already famous for its Explorer's Inn Reserve and other research and ecotourism camps on the lower part of the river, the area to the east and

esta actividad económica todavía no ha dado muchas ganancias a los productores de la región. El entendimiento de todas estas variables socioeconómicas determinaron que los talleres de planificación de conservacion en Puno fueran diferentes a los de Madre de Dios. La respuesta de la gente local ha sido favorable hasta ahora, porque sus propias experiencias les han enseñado que de la capacidad de ellos principalmente y no de subsidios externos, dependen las soluciones a los problemas en el uso de los recursos.

El proceso del planeamiento conservacionista en Tambopata no está todavía terminado. No es suficiente tener el parque nacional declarado. Los conservacionistas deben trabajar muy cerca con la población local para poder encontrar alternativas viables tanto económicas, sociales y biológicas; que

south of the Río Tambopata is mostly a large and diverse wilderness, encompassing many distinct habitats of the lower ridges of the Andes and upper Amazonian plain. It has had little recent human disturbance and is of enormous conservation importance especially in context with the adjacent Río Madidi region of Bolivia and the Manu region to the north.

2 The Río Heath and Pampas del Heath National Sanctuary. The Peruvian frontier with Bolivia already has a designated National Sanctuary to protect the distinctive pampas vegetation in Perú. However, the reserve boundaries and its management have not fully taken into account the importance of the context and dynamics of the Río Heath ecosystems (quite different from those of the rest of Perú) of which it is a part and which are crucial for its long-term survival.

Both areas offer unique opportunities for long-term research projects on their biological communities. The Explorer's Inn Reserve has excellent accommodation facilities and is only a few hours away from the airport at Puerto Maldonado. The Ccolpa de Guacamayos and the Pampas del Heath Sanctuary have basic facilities which, with a modest additional investment, could provide adequate support to researchers.

Río Tambopata

The vegetation adjacent to the Río Tambopata, which flows from steep mountain slopes, through low foothills and onto flat terraces of the Amazon plain, is summarized below for each of our major field sites. The composition of the flora and fauna recorded here is not unexpected for the southwestern part of the Upper Amazon drainage. What is distinctive about the area is the diversity of habitats, which in turn leads to the cumulative richness in species.

What is distinctive about the area is the diversity of habitats, which in turn leads to the cumulative richness in species.

permitan la protección de la biodiversidad y el uso sostenido de los recursos. Este es el reto que nosotros debemos encontrar si queremos que la riqueza biológica de Tambopata-Candamo se mantenga.

RESÚMEN DE RESULTADOS

Hemos enfocado en dos áreas de importancia conservacionista principales que estan contiguas en la base de los Andes al sudeste peruano, adyacentes al bosque menos perturbado de Bolivia, el cual ha sido propuesto para ser incuido dentro del Parque Nacional Madidi.

1 El Río Tambopata y los Cerros del Távara. Aunque ya famoso por la Reserva del Explorer's Inn, por otras investigaciones, así como por el ecoturismo en la parte baja del río, el área al este y sur del Río Tambopata es mayormente un área silvestre grande y diversa, comprendiendo varios habitats característicos de las partes bajas de la Cordillera de los Andes y llanos de la alta Amazonía. Ambas áreas han sufrido pequeñas perturbaciones humanas recientes, y su conservación es considerada de enorme importancia porque se encuentran adyacentes al Río Madidi en la región boliviana y a la región del Manu al norte.

2 El Río Heath y El Santuario Nacional Pampas del Heath. La frontera peruana con Bolivia ya tiene designado al Santuario Nacional para proteger la peculiar vegetación de las pampas en Perú. Sin embargo, los límites de la reserva y su manejo no han tomado en cuenta la importancia del contexto y dinámica de los ecosistemas del Río Heath (totalmente diferente a aquellos en el resto del Perú) del cual este es parte, y que son cruciales para su sobrevivencia a largo plazo.

Ambas áreas ofrecen oportunidades únicas para proyectos de investigación a largo

We recorded 72 species of mammals in 19 days in the upper Río Tambopata-Távara area. Adding others previously known from this area and the Explorer's Inn Reserve brings to 103 species the mammals now known from the Río Tambopata area as a whole, with other species expected. In addition to numerous records of large mammals, our preliminary results show that the Tambopata basin has a high diversity of smaller mammals, such as rodents and bats. The mosaic habitat is a major source of the high species diversity of mammals.

Of the 107 frog species now known from the Madre de Dios area, six (5.6% of the total) were recorded for the first time during the RAP expedition and a total of 13 species (15.6%) are new reports for the Tambopata-Candamo Reserved Zone. The richness in species of other animal groups is similarly high in the Tambopata region as summarized below for specific areas.

There are high populations of species that have become rare from overhunting in many parts of Amazonia, especially tapirs and spider monkeys, but also jaguars, capybaras, white-lipped peccaries, mid-sized and large monkeys, caimans and river turtles. The river supports at least several groups of giant otters, as well as river otters. Woolly monkeys are uncommon or rare over most of their range and are threatened or extinct from hunting in many regions; however, groups of this species occur north of the Río Tambopata in the foothills near the boca Río Távara (W. Wust, pers. comm.). The presence of large individuals and sizeable populations of many of these species in the area indicates low human hunting pressure. The upper Tambopata drainage provides a renewable source of wildlife used by residents of the lower river and is thus important for maintaining the basic resources used by the local human communities. The high price of gaso-

plazo en sus comunidades biológicas. La Reserva del Explorer's Inn cuenta con excelentes facilidades y está solamente a pocas horas del aeropuerto de Puerto Maldonado. La Ccolpa de Guacamayos y el Santuario de las Pampas del Heath cuentan con las instalaciones básicas, que con un poco de inversión adicional podrían proveer un apoyo adecuado a los investigadores.

Río Tambopata

La vegetación adyacente al Río Tambopata, el cual fluye desde las pendientes acentuadas por las zonas bajas del pie de montaña y hacia las terrazas planas del llano Amazónico, está descrita más adelante para cada uno de nuestros principales sitios de estudio. La composición de la flora y fauna registradas aquí no es inesperada para la parte sudoeste del drenaje del Alto Amazonas. Lo que es peculiar del área es la diversidad de habitats, que conlleva a una cumulativa abundancia en especies.

Registramos 72 especies de mamíferos en 19 días en el área alta del Río Tambopata-Távara. Adicionando otras previamente conocidas de esta área y de la Reserva Explorer's Inn se llega al número de 103 especies de mamíferos conocidas ahora para toda el área del Río Tambopata. Además de numerosos registros de mamíferos grandes, nuestros resultados preliminares mostraron que la cuenca del Tambopata tiene una alta diversidad de pequeños mamíferos, tales como roedores y murciélagos. La principal fuente de esta alta diversidad en especies es la cantidad de habitats que se encuentran.

De las 107 especies de ranas conocidas hasta ahora para el área de Madre de Dios, seis (5.6% del total) fueron registradas por primera vez durante la expedición del RAP y un total de 13 especies (15.6%) son nuevos reportes para la Zona Reservada de

Lo que es peculiar del área es la diversidad de habitats, que conlleva a una cumulativa abundancia en especies.

line is now the chief curb on subsistence and market hunting far upriver from villages, but this situation may not last indefinitely.

Cerros del Távara

The vegetation in the low mountains reflects the underlying geological differences, with the greatest contrast in flora found between the quartzite-based sandy clay soils and the pure clay soils. The differences in the forests between two of the geologically distinct Távara ridges is considerable, with the great majority of species that are common on one ridge being rare or absent on the other.

Altitudinal differences in the flora are much less striking. The plant communities at the base of the ridge, 300 m, and near the crest at 800-900 m are not significantly different. Many of the species on the high rocky floodplain are also found at the highest elevations. One can see *Dipteryx micrantha*, a characteristic emergent floodplain tree, growing way upslope into the mossiest forest on the ridge-crest at 825 m. There are very few cloud forest species among the terrestrial woody plants. On the other hand, one team member collected 80 species of orchids, many of which are probably more abundant in cloud forests of higher elevations. As a whole the Cerros del Távara has a very mixed forest in comparison with the lowlands, with considerable difference from hectare to hectare in composition.

Of particular ecological interest are the occasional emergent gymnosperm trees, *Podocarpus* sp., between 700 and 800 m; the large number of species occurring in discrete clumps (nonclonal) with an explosive dispersal mechanism; and the frequency of "chacras del diablo," distinctive understory clearings with high densities of ant-associated plants. The Cañón del Távara, a narrow gorge, has a distinctive flora associated with its steep rocky banks.

The differences in the forests between two of the geologically distinct Távara ridges is considerable ....

Tambopata-Candamo. La riqueza en especies de otros grupos de animales es similarmente alta en la región de Tambopata como es resumido más adelante para áreas específicas.

Hemos encontrado que hay poblaciones grandes de especies que en muchas partes de la Amazonía ya son raras debido a la sobrecaza, especialmente tapires y monos araña, pero también para jaguares, capibaras, huanganas, monos medianos y grandes, caimanes y tortugas de río. En los ríos se encuentran al menos varios grupos de Lobos de río, así como nutrias. Los monos choro son raros o poco comunes y están amenazados o extintos por la caza en muchas regiones; sin embargo, grupos de esta especie viven al norte del Río Tambopata, al pie de la montaña cerca a la boca del Río Távara (W. Wust, com.pers.). La presencia de ejemplares grandes, y de grandes poblaciones de muchas de estas especies en el área indican la baja presión de caza por humanos.

La cuenca del alto Tambopata provee una fuente renovable de fauna silvestre usada por los residentes de la parte baja del río y es por lo tanto importante para el mantenimiento de los recursos básicos usados por las comunidades humanas locales. El alto precio de la gasolina es ahora lo que determina la baja presión de caza para subsistencia y el mercadeo río arriba de los pueblos, pero esta situación no puede durar indefinidamente.

Cerros del Távara

La vegetación en las montañas bajas reflejan lo marcado de las diferencias geológicas, con un gran contraste entre la flora encontrada en los suelos arenosos-arcillosos y los puramente arcillosos. La diferencia que existe entre los bosques de ambas laderas de los Cerros del Távara es de considerar, mientras que a un lado la gran mayoría de especies son comunes al otro son raros o no se encuentran.

Although the composition of plant families in the Távara forest is similar to that of adjacent lowland Amazonia, the species composition is completely different (including unusually high numbers of the family Melastomataceae). With respect to conservation, this is the most important aspect of the foothill forest in the Tambopata-Candamo area: it is a completely different set of taxa than are represented in the lowlands away from the mountain base.

The birds on this ridge are dominated by two elements, Amazonian species that are at or nearing their upper elevational limit, and species largely or entirely restricted to a narrow band of moist forest on the lowermost slopes of the Andes. We recorded an impressive 38 bird species in the latter category in spite of the Cerros del Távara being at or near the lower elevational range of many of these species. Due to the elevational restriction, and the high rate of forest clearance within this zone, this lower montane zone is one of the most critically threatened forest habitats in all of South America. Several of the bird species encountered are extremely rare and vulnerable to extinction.

Among the 24 species of frogs recorded, the fauna up to 500 m elevation was similar to that found in the Tambopata lowlands away from the Andes. But above that elevation, several different species were recorded including a new species of poison dart frog (*Epipedobates*) now being described. Some of the ridge species, including two undescribed species of *Eleutherodactylus*, are known from the lowlands to the north in Manu, but not the Tambopata lowlands.

Ccolpa de Guacamayos

Dense stands of two large bamboo species cover much of the hilly area, from which the clay-lick is exposed, and the surrounding allu-

Las diferencias altitudinales son mucho menos notables en la flora. Las comunidades de plantas en la base de los cerros, 300 m. y cerca de la cresta a 800-900 m. no son significativamente diferentes. Muchas de las especies de las altas llanuras aluviales rocosas son también encontradas en las elevaciones más altas. Se puede ver *Dipteryx micrantha*, un árbol característico del dosel de las llanuras, creciendo hacia arriba de la ladera hasta la cresta, entre el bosque musgoso a 825 m. Hay muy pocas especies de bosque de neblina entre las plantas terrestres leñosas. Por otro lado, un miembro del equipo colectó 80 especies de orquídeas, muchas de las cuales son probablemente más abundantes en bosques de neblina a altitudes mayores. En conjunto, los Cerros del Távara presentan un bosque muy mezclado en comparación con las tierras bajas, con una diferencia considerable en composición de hectárea a hectárea.

De interés ecológico particular son las ocasionales apariciones de gimnospermas, *Podocarpus* sp., entre los 700 y 800 m.; el gran número de especies que ocurren en discretos grupos (noclonal) con un mecanismo explosivo de dispersión; y la frecuencia de las "chacras del diablo", características zonas casi desprovistas de vegetación baja, con alta densidad de plantas asociadas a hormigas. El Cañón del Távara, una garganta estrecha, tiene una peculiar flora asociada con sus empinadas laderas rocosas.

Aunque la composición de las familias de plantas en los bosques del Távara es similar a aquellos de las tierras bajas de la Amazonía adyacente, la composición de especies es completamente diferente (incluyendo la inusual presencia de grandes números de plantas de la familia Melastomataceae). Con respecto a la conservación, éste es el aspecto más importante del bosque de pie de montaña en el área del Tambopata-Candamo: es una composición taxonómica completamente

La diferencia que existe entre los bosques de ambas laderas de los Cerros del Távara es de considerar ....

vium. The origin of these large bamboo stands, which are common below the foothills of the Andes in this region, remain a mystery even though their impact is very obvious. Bamboo now actively colonizes any abandoned human clearing and areas subject to human intervention, e.g., areas with logging activity. Bamboo, in limited quantities, has some commercial value and is the unique habitat of several bird and mammal species. But it appears that any extensive logging activity in this region will result in a takeover by bamboo which, once established, is extremely difficult to eradicate and inhibits the development of most commercially valuable forest species as well as severely reducing the biological diversity.

There are beautiful stands of typical old-growth floodplain forest near the river along this area of the Tambopata. This kind of forest, dominated by several giant tree species such as *Ceiba pentandra*, is nearly wiped out in the Amazon basin because it is the forest type that is most accessible from the river "highways" and is most likely to be colonized for crop cultivation. The other vegetation near the camp includes different successional series typical of braided rivers and gravelly substrate, but with higher densities of trunk-climbers and epiphytes associated with greater year-round moisture. Among the bamboo on the hills, especially along stream bottoms, are patches of typical low hill-forest. At the base of the hills are occasional *Mauritia* palm swamps mixed with *Lueheopsis*, constituting the northern limit of what is a typical association in much of the forest swamps of Bolivia.

As one would expect, the flora and fauna of the Ccolpa is very similar to that of the Explorer's Inn Reserve farther downstream. The Ccolpa's closer proximity to the Andean foothills is reflected in the presence of, or great abundance of, primarily foothill species.

One of the most notable features of the

There are beautiful stands of typical old-growth floodplain forest near the river along this area of the Tambopata.

diferente a lo que está representado en otras tierras bajas lejos de la base de la montaña.

Las composición de las aves en este cerro están dominadas por dos elementos: las especies amazónicas que están o se encuentran cerca de sus límites superiores de elevación; y especies que están total o parcialmente restringidas a una banda muy estrecha del bosque húmedo en las pendientes más bajas de los Andes. Registramos el impresionante número de 38 especies de aves en esta última categoría a pesar de que los Cerros del Távara, estaban para muchas de estas especies cerca, o al límite de su rango de elevación más bajo. Debido a las restricciones altitudinales, y la alta tasa de deforestación en estos límites altitudinales, esta zona montano bajo es uno de los habitats más amenazados en toda America del Sur. Muchas de las especies de aves encontradas son extremadamente raras o vulnerables a extinción.

Entre las 24 especies de ranas registradas, la fauna hasta 500 m. de elevación fue similar a aquella encontrada en las tierras bajas de Tambopata lejos de los Andes. Pero sobre esa elevación, muchas especies diferentes fueron registradas incluyendo una nueva especie de rana venenosa (*Epipedobates*) que ahora esta siendo descrita. Algunas de las especies de la cresta del cerro, incluyendo dos especies no descritas de *Eleutherodactylus*, son conocidas de las tierras bajas al norte en Manu, pero no de las del Tambopata.

Ccolpa de Guacamayos

Grupos densos de bambús de dos especies cubren mucho de esta zona de topografía ondulada, desde la cual se expone la Ccolpa, y la circundante tierra aluvial. El origen de estos grandes bambús, que son comunes en la parte baja de los Andes en esta región, todavía es un misterio aun cuando su impacto es muy obvio. El bambú coloniza cualquier claro o

Ccolpa, of course, is the large number of parrots and macaws that visit the clay lick. Also notable are the large populations of birds that are closely associated with bamboo, including what may be the largest population density of the extremely scarce antbird *Formicarius rufifrons* (Kratter, in press), whose known distribution lies entirely within the Department of Madre de Dios.

In only eight days of sampling at the Ccolpa, we recorded a total of 44 anuran species, 12 lizards, and ten snakes.

Explorer's Inn Reserve (Tambopata Reserved Zone)

This major center of tropical forest research and ecotourism has an extraordinary diversity of old and new terraces, oxbow lakes, swamps and other habitats. Any single one of these habitats may not have unusually high numbers of species associated with it, but taken as a whole, the area of the reserve is extremely rich in species. It has a good representation of many of the habitats or vegetation types of the region, though by no means all. Some of its habitats, such as the very high, but recent, clay terraces are rare in this region.

The flora in the reserve (ca. 1400 species of vascular plants) is fairly typical of the southwestern Amazon Basin. The permanent forest plots that have been established in the reserve have been instrumental in revealing many new species to science and new records for Perú, in suggesting hypotheses about the effects of global climate change on tropical forest turnover rates, and in providing a basis for other kinds of research.

Almost 575 bird species have been recorded at the Explorer's Inn Reserve, all within an area of approximately 5000 ha, an extraordinarily high total explained largely by the habitat diversity. Almost 50% of the forest bird species here are more or less restricted to

lugar abandonado después de la intervención humana, por ejemplo, áreas de actividad maderera. El bambú, en cantidades limitadas, tiene algún valor comercial y es el único habitat de muchas especies de aves y mamíferos. Pero parece que cualquier extensa actividad maderera en esta región resultaría en una sobrepoblación de bambú que, una vez establecida, es extremadamente difícil de erradicar e inhibe el desarrollo de muchas especies forestales valiosas así como tambien reduce severamente la diversidad biológica.

Cerca al río a lo largo de esta área de Tambopata se encuentran hermosas áreas de típicos bosques maduros de llanos inundados. Este tipo de bosque, dominado por muchas especies arbóreas gigantes tales como *Ceiba pentandra*, ha sido casi erradicado de la cuenca del Amazonas porque este es el tipo de bosque que tiene fácil acceso por las "carreteras fluviales" (ríos) y que es el mas probable de ser colonizado para campos de cultivo. La otra vegetación cerca al campamento incluye diferentes series sucesionales típicas de ríos entrelazados y sustrato pedregoso, pero con densidades mas altas de plantas "trepadoras" y epífitas asociadas a una mayor humedad que dura todo el año. En medio del bambú de las montañas, especialmente a lo largo del fondo de las quebradas, se encuentran manchas de bosques típicos de montañas bajas. En la base de las montañas se encuentran ocasionales pantanos con palmeras de Aguaje (*Mauritia flexuosa*) mezclados con *Lueheopsis*, constituyendo el límite norte de lo que es una típica asociación en muchos de los bosques pantanosos de Bolivia.

Tal como era de esperar, la flora y fauna de la Ccolpa es muy similar a la que presenta la Reserva del Explorer's Inn río abajo. La proximidad de la Ccolpa al pie de los Andes es reflejada en la presencia de, o la gran abundancia de, especies primariamente presentes al pie de montaña.

Cerca al río a lo largo de esta área de Tambopata se encuentran hermosas áreas de típicos bosques maduros de llanos inundados.

the vegetation on different river terraces subject to flooding; less than 10% of the forest bird species are similarly restricted to *terra firme* forests on higher ground that never experience flooding. Approximately 80 species recorded at the Reserve are known or strongly suspected to be migrants to Amazonia. Of those species whose geographic origin can be inferred, equal numbers are of austral (37) as opposed to north temperate (35) origin. Approximately two-thirds of the bird species endemic to the lowlands of southwestern Amazonia are found at this reserve. Seven of these bird species have been described as new to science within the past 30 years. Although macaws are still relatively common at the reserve, populations of some other large species, such as Razor-billed Curassow (*Mitu tuberosa*), are very low, probably a reflection of past hunting pressure.

The small area of the Explorer's Inn Reserve shelters over 1200 species of butterflies, the second-richest documented butterfly community in the world (Manu being first). This fact alone underlines the extraordinary importance of conserving the area. At least one-third of the butterfly fauna of Perú has been recorded at two small sites in Madre de Dios (Explorer's Inn and Pakitza Station in Manu), highlighting the critical diversity of the area. In ten effective surveying days spent at Tambopata as part of the RAP trip, at least seven new species were recorded for the area, despite the fact that butterfly populations were very low as a result of unfavorable climatic conditions.

Río Heath and Pampas del Heath

The Río Heath and its pampas represent an ecosystem unique in Perú; the Peruvian pampas are but a fraction of the size of extensive pampas across the border in Bolivia. The Río Heath itself not only provides access to the

Una de las más notables características de la Ccolpa, por supuesto, es el gran número de loros y guacamayos que visitan la ladera con alto contenido de sales y minerales. Es también notable la gran población de aves que están asociadas con el bambú, incluyendo lo que podría ser las más grande y densa población de la rara ave seguidora de hormigas *Formicarius rufifrons* (Kratter, en prensa), de la cual su distribución total conocida está dentro del Departamento de Madre de Dios.

En solamente ocho días de muestreo en el Ccolpa, registramos un total de 44 especies de anuros, 12 de lagartijas, y 10 de serpientes.

Reserva del Explorer's Inn (Zona Reservada Tambopata)

Este centro principal de investigación de bosque tropical y ecoturismo tiene una extraordinaria diversidad de viejas y nuevas terrazas, cochas, pantanos y otros habitats. Cualquiera de estos habitats por si solo puede que no tenga un inusual número de especies asociadas a éste, pero tomándola en cuenta como un todo, el área de la reserva es extremadamente rica en especies. Cuenta con una buena representación de muchos de los habitats o vegetación tipo de la región, lo que no significa de todos. Algunos de sus habitats, tales como las muy altas, pero recientes, terrazas de arcilla son raras en esta región.

La flora en la reserva (+/-1400 especies de plantas vasculares) es bastante típica de la cuenca sudoeste del Amazonas. Las parcelas permanentes de bosques que fueron establecidas en la reserva varios años atrás, han sido instrumentales para el descubrimiento de varias especies nuevas para la ciencia, y nuevos registros para el Perú, ademas de sugerir nuevas hipótesis acerca de los efectos del cambio global del clima en los bosques tropicales, y en proveer nuevas bases para otros tipos de investigación.

famed Pampas del Heath, but includes a great extension of unusual clay terraces back from the river called *sartenjales* or *shebonales*. Like the pampas, these terraces, at least well-developed ones, seem to be rare in the rest of Amazonian Perú, though more frequent to the south in Bolivia. There appears to be a connection between these terraces and the exposed clay licks or ccolpas that attract congregations of macaws and other animals, but the generality of this association remains to be determined.

Pampas are periodically inundated grasslands maintained by occasional burning in the dry season. They were previously much more extensive in Perú to the north of the current pampas, and have mostly reverted to forest after the cessation of burning. The origin and age of the pampas is not known, but their future and the future of the organisms associated with them depends on periodic fires.

The pampas are perched on a high terrace, underlain by an impermeable clay. Small pockets of forest develop on the areas that are either the driest or wettest, the termite mounds or the wet depressions. The flatter areas are dominated by grasses, sedges, and other herbs, with a scattering of shrubs, mostly in the Melastomataceae. Many of the plant species recently collected here are new records for Perú. The forests that regenerate on unburned pampa are barely known, but our trip revealed that some of them are nearly solid stands of a single tree species, *Qualea wittrockii*.

Fourteen species of birds are known in Perú only from the Pampas del Heath, and several additional species of the Pampas are recorded from only two or three other sites within Perú.

We recorded a total of 68 species of mammals in the region of Pampas del Heath National Sanctuary, 15 of these in the open

Casi 575 especies de aves han sido registradas en la Reserva del Explorer's Inn, dentro de un área de aproximadamente 5000 ha., un total elevado que es explicado ampliamente por la diversidad de habitats. Casi el 50% de las especies de aves del bosque aquí están más o menos restringidas a la vegetación en las diferentes terrazas sujetas a inundaciones; menos del 10% de las especies de aves están similarmente restringidas a bosques de "tierra firme" en los terrenos altos que nunca sufren inundaciones. Aproximadamente 80 especies registradas en la Reserva son conocidas (o se sospecha) que son migratorias a la Amazonía. De aquellas especies de quienes su origen puede ser inferido, en igual número son australes (37), como del norte temperado (35). Aproximadamente dos tercios de las especies de aves endémicas de tierras bajas del sudoeste de la Amazonía son encontradas en esta reserva. Siete de estas especies de aves han sido descritas como nuevas para la ciencia en los últimos 30 años. Aunque los guacamayos son relativamente comunes en la reserva, poblaciones de otras especies grandes, tales como el paujil (*Mitu tuberosa*), son muy bajas, probablemente como reflejo de la sobrecaza en el pasado.

El área pequeña de la Reserva del Explorer's Inn alberga mas de 1200 especies de mariposas, la segunda comunidad más rica documentada en el mundo (Manu es la primera). Este solo hecho determina la extraordinaria importancia de conservar el área. Por lo menos un tercio de la fauna lepidóptera del Perú ha sido registrada en dos pequeños lugares en Madre de Dios (Explorer's Inn y la Estación Pakitza en Manu), remarcando la extraordinaria diversidad del área. Durante 10 días de efectivo muestreo hechos en Tambopata como parte del viaje del RAP, al menos siete nuevas especies fueron registradas para el área, a

Casi 575 especies de aves han sido registradas en la Reserva del Explorer's Inn ....

pampa, and 52 in the closed-canopy forest or on the riverbank. Another six species were found by earlier expeditions, bringing the total recorded for the Sanctuary to 74 species. This includes the globally important marsh deer, and the maned wolf and giant anteater reported by an earlier expedition. Important sightings of rare mammals on the RAP trip included a pacarana (rodent) and a short-eared dog. We can confirm the presence of four mammals that are strictly pampa species in the Sanctuary, and a fifth has been reconfirmed. Two others need to be verified. Of the five, four occur nowhere else in Perú. The mammal fauna is thus small, but significant. The pampas may recently have shrunk to below the size needed for a breeding population of maned wolves, although the Río Heath may not be much of a barrier to dispersal from the much more extensive pampas in Bolivia.

In this area, we recorded 28 species of amphibians and 17 species of reptiles. A collecting survey of the freshwater fishes of six water bodies in the drainage of the Río Heath, recorded 95 species. The fish diversity is related to the many different aquatic habitats found in the Sanctuary.

For at least 29 species of butterflies newly recorded for Perú, the Pampas del Heath represent their westernmost geographical extension, and it is unlikely that most of them will be found elsewhere in Perú. The forest surrounding the savannas at Pampas del Heath National Sanctuary might be as rich in butterfly species as that of Explorer's Inn (which is only 50 km away), with the added bonus of harboring savanna species not found there.

CONSERVATION OPPORTUNITIES

Portions of the following discussion of conservation, management, and research opportunities have previously been presented in public meetings by various combinations of the

Important sightings of rare mammals on the RAP trip included a pacarana (rodent) and a short-eared dog.

pesar del hecho de que las poblaciones de mariposas estaban muy bajas como resultado de las condiciones climáticas desfavorables.

Río Heath and Pampas del Heath

El Río Heath y sus pampas son un ecosistema único en el Perú; las pampas peruanas son solo una fracción de las extensas pampas al otro lado de la frontera en Bolivia. El Río Heath por si mismo no solo provee acceso a las famosas Pampas del Heath, sino que incluye una gran extensión de las inusuales terrazas arcillosas que vienen desde el río, llamadas sartenjales o shebonales . Como las pampas, esta terrazas, por lo menos las bien desarrolladas, parecen ser raras en el resto de la Amazonía Peruana, aunque son vistas más frecuentemente al sur en Bolivia. Parece ser que hay una conección entre estas terrazas y las Ccolpas que atraen congregaciones de guacamayos y otros animales, pero la generalidad de esta asociación espera ser determinada.

Las Pampas son pastizales inundados periodicamente, mantenidas por ocasionales incendios en la temporada seca. En el pasado, ellas fueron mucho más extensas en el Perú hacia el norte de las actuales pampas, mayormente ahora revertidas en bosques después que los incendios cesaron. El origen y edad de las pampas no es conocido, pero su futuro y el futuro de los organismos asociados a ellas dependen de los periódicos fuegos.

Las pampas están asentadas en terrazas altas, sostenidas por arcilla impermeable. Pequeños bosques se desarrollan en áreas que son o muy secas o muy húmedas: los termiteros o las depresiones húmedas. Las áreas mas planas son dominadas por pastos, juncos, y otras hierbas, con algunos arbustos desperdigados, principalmente Melastomataceae. Muchas de las especies de plantas recientemente colectadas aquí son nuevos registros

RAP researchers. Preliminary observations made by the RAP expeditions were presented in a workshop at Explorer's Inn on 20 and 21 June, 1992. Core team members present were Parker, Emmons, and Foster. Nuñez presented Gentry's preliminary results. Peruvian counterpart team members also presented their observations; i.e., Ortega, Icochea, Ascorra, Romo, Lamas, and Rodríguez. This meeting was tape recorded and later transcribed and used by the CI-Perú staff in discussions and proposals concerning the fate of the Tambopata-Candamo Reserved Zone. There was also a public presentation by the RAP team in Puerto Maldonado on the 22nd of June, 1992 that was attended by Parker and the Peruvian counterparts. A year later, on July 4-5, 1993, there was a public forum held in Puerto Maldonado on a zoning proposal for the TCRZ. Parker and Chicchón attended to help insure that the findings of the RAP survey were included in the planning process for the protected area.

The net result of this participatory process is that some of the following recommendations have already entered into the plans for the permanent zonation of the Tambopata-Candamo Reserved Zone, including the proposed Bahuaja-Sonene National Park (INRENA 1994). The following comments on the boundaries for a national park were drafted by Ted Parker, to which are added suggestions for conservation, economic development, management, and research in the Tambopata-Heath area of southeastern Perú proposed by the research team collectively.

1 Comments on the establishment of a national park within the Tambopata-Candamo Reserved Zone. There are a number of distinct types of tall evergreen forest on the floodplains of the Río Tambopata in the lowlands between the mouth of the Río Távara and the mouth of the Tambopata itself at

para el Perú. Los bosques que se regeneran en las pampas no quemadas son muy poco conocidos, pero nuestro viaje reveló que algunos de ellos son sólidas estaciones de una sola especie arbórea, *Qualea wittrockii*.

En el Perú catorce especies de aves son conocidas solamente de las Pampas del Heath, y muchas otras son registradas en uno o dos lugares más en el Perú.

Hemos registrado un total de 68 especies de mamíferos en la región del Santuario Nacional de las Pampas del Heath, 15 de estos en pampa abierta, y 52 en el bosque de dosel contínuo o en la orilla del río. Otras seis especies fueron encontradas por expediciones anteriores, dando un total de 74 especies para el santuario. Estas incluyen al mundialmente importante ciervo de los pantanos, al lobo de crín y al oso hormiguero gigante. Reportes importantes de mamíferos raros en el viaje del RAP incluyen la pacarana (roedor) y un perro de monte de orejas cortas. Pudimos confirmar la presencia de cuatro mamíferos que son estrictamente especies de las pampas en el Santuario Nacional, y un quinto ha sido reconfirmado. Otros dos necesitan ser verificados. De los cinco, cuatro no ocurren en otro lugar en el Perú. La fauna mamífera es tal vez pequeña, pero significativa. Recientemente las pampas podrían haberse reducido a un tamaño menor al necesario para mantener una población del lobo de crín, aunque el Río Heath puede que no sea una barrera para la dispersión de éste desde las más extensas pampas en Bolivia.

En esta área, registramos 28 especies de anfibios y 17 especies de reptiles. En un muestreo del Río Heath en seis cuerpos de agua en el Santuario Nacional de las Pampas del Heath, registramos 95 especies de peces. La diversidad de ellos esta relacionada a los diferentes habitats acuáticos encontrados en el santuario.

Para al menos 29 especies de mariposas nuevas registradas para el Perú, las Pampas del

Reportes importantes de mamíferos raros en el viaje del RAP incluyen la pacarana (roedor) y un perro de monte de orejas cortas.

In the upper Tambopata region, we see a clear and perhaps unique opportunity to protect one of the few remaining uninhabited sections of whitewater river in lowland eastern Perú.

Puerto Maldonado. These forests are biologically important because they are exceptionally species rich (including, for example, large numbers of economically valuable plant species), and they also support, in the case of birds at least, nearly all of the species confined to the lowlands of southwest Amazonia. Differences in the plant, bird, and possibly even amphibian communities distinguish these forest types. Such community differences undoubtedly reflect variation in local rainfall, flooding, alluvial deposition, degree of human and natural disturbance, and many other factors. The conservation significance of this is clear: in order to maintain the full-range of riverine ecosystems and species in southeastern Perú, a large section (or sections) of the Río Tambopata should be included in a conservation unit of some type, preferably as part of the proposed National Park. At the very least, human colonization in some of these biologically important floodplain areas should be limited or prohibited altogether. As a compromise of last resort, low-level, carefully monitored extraction of forest-based plant and animal products could be allowed, especially if these activities resulted in direct benefits to the local communities downriver. In the upper Tambopata region, we see a clear and perhaps unique opportunity to protect one of the few remaining uninhabited sections of whitewater river in lowland eastern Perú. Specifically, based on both the biological argument made above (and other considerations discussed below), we recommend the inclusion of that portion of the river which lies between the mouth of the Río Távara and the mouth of the Río Malinowski. Such action appears to us to be critical to the long-term success of any major conservation initiative in the region. Failure to protect forests along this portion of the Tambopata will undoubtedly be followed over the next 5-10 years by human colonization of this area, which in turn

Heath representan su límite oeste en su distribución geográfica, y no parece ser que muchas de ellas se encuentren en otros lugares del Perú. Los bosques que rodean las savanas en el Santuario Nacional de las Pampas del Heath podrían ser tan ricos en especies de mariposas como el Explorer's Inn (que está solamente a 50 Km), adicionándole además especies de la savana que no se encuentran allí.

OPORTUNIDADES DE CONSERVACIÓN

Partes de la siguiente discusión acerca de las oportunidades de conservación, manejo, e investigación han sido previamente presentadas en reuniones públicas por varias combinaciones de los investigadores del RAP. Las observaciones preliminares hechas por el RAP fueron presentadas en un taller en el Explorer's Inn en Junio 20 y 21 de 1992. Miembros principales del equipo presentes fueron Parker, Emmons, y Foster. Nuñez presentó los resultados preliminares de Gentry. El equipo de la contraparte peruana también presentó sus observaciones; entre otros: Ortega, Icochea, Ascorra, Romo, Lamas, y Rodríguez. Esta reunión fue grabada y más tarde transcrita y usada por el personal del CI-Perú en discusiones y propuestas concernientes al futuro de la Zona Reservada Tambopata-Candamo. También hubo una presentación pública por el grupo del RAP en Puerto Maldonado en Junio 22 de 1992 en la que participaron Parker y la contraparte peruana. Un año después, en Julio 4 y 5 de 1992, se llevó a cabo un foro público en Puerto Maldonado sobre la propuesta de la Zona Reservada Tambopata-Candamo. Parker y Chicchón estuvieron presentes para asegurar que lo encontrado por el RAP fuera incluído en el proceso de planificación del área protegida.

El resultado neto de este proceso participatorio es que algunas de las siguientes

would result in ever-increasing pressure on the so-called core area of the national park in the foothills above 400 m.

An additional argument that can easily be made for including the Tambopata above the Malinowski in the Park derives from our sincere belief that local people, especially those who live near the mouth of the Malinowski and downriver for many kilometers, could ultimately benefit more from the development of ecotourism along the upper Tambopata than they will from present forms of agriculture or timber exploitation. The scenically beautiful Ccolpa de Guacamayos area, which attracts immense numbers of macaws and other parrots, can without any doubt become one of the premier destinations for natural history tour groups in South America. This attraction, combined with the allure of the region's existing tourist facilities such as Explorer's Inn, Cuzco Amazonico, and Manu Lodge, could and should become the focal points for revitalizing a once prosperous industry in southeast Perú. If this occurs, we can only hope that everyone involved in that business will seek creative and appropriate ways of involving local people in their efforts. Certainly such an industry, which could over time attract thousands of tourists, would result in a much greater and more secure form of economic benefit than that presently derived from shifting subsistence agriculture, non-sustainable timber extraction, or gold mining.

Strong consideration should be given to the inclusion in the Park of representative examples of floodplain and upland forests on high alluvial terraces along the lower Tambopata in the vicinity of the Explorer's Inn. By working (to some degree) with local people, the management of this tourist facility has been able, over the course of 19 years, to protect much of the riverine forest between their property at the mouth of the Río La Torre, and the nearest Amerindian commu-

recomendaciones ya están contempladas dentro de los planes para la permanente zonificación de la Zona Reservada Tambopata-Candamo, incluyendo el propuesto Parque Nacional Bahuaja-Sonene (INRENA 1994). Los comentarios siguientes sobre los límites para el parque nacional fueron delineados por Ted Parker, a los cuales se adiciona las sugerencias para la conservación, desarrollo económico, manejo, e investigación en el área sudeste de Tambopata-Heath Perú propuestas colectivamente por el equipo de inves-tigadores.

1 Comentarios sobre el establecimiento de un Parque Nacional dentro de la Zona Reservada Tambopata-Candamo. Hay varios tipos distintos de bosques siempre-verdes de gran altura en los llanos inundados del Rio Tambopata en las tierras bajas entre la boca del Río Távara y la boca del mismo Tambopata en Puerto Maldonado. Estos bosques son biológicamente importantes porque son excepcionalmente ricos en especies (que incluyen por ejemplo, a un gran número de plantas de gran valor económico), y ellos también albergan en el caso de las aves por lo menos, a cerca de todas las especies confinadas a las tierras bajas al sudoeste de la Amazonía. Las diferencias en las comunidades de plantas, aves, y posiblemente anfibios distinguen estos tipos de bosques. Tales diferencias en las comunidades sin duda reflejan las variaciones locales de las lluvias, inundaciones, deposiciones aluviales, y grado de perturbación natural y humana, y muchos otros factores. La significancia conservacionista de esta área es clara: para poder mantener una muestra completa de los ecosistemas rivereños y especies al sudeste peruano, una gran sección (o secciones) del Rio Tambopata debería ser incluido en una unidad de conservación, preferiblemente como parte del propuesto Parque Nacional. Por lo menos, la coloniz-

nity well down the river at Infierno. Although the relationship between those at the Inn and those in the village is complicated and often strained, the end result of this continuous contact has been the survival of forests and wildlife—and the potential they represent for research and ecotourism. In contrast, nearly all of the forest bordering the Tambopata above the La Torre (all the way to the Malinowski), and below it to Puerto Maldonado, has been replaced by agriculture and cattle ranches. Because of its well-documented biological importance, as well as its strategic position at the mouth of the La Torre (which flows out of a large uninhabited area adjoining the lower Río Heath drainage), and because of the long history of ecological research in the surrounding forests, the Explorer's Inn deserves special consideration as a base of operations for any type of conservation area established in the region. It already serves numerous purposes that include: (1) its operation as a tourist facility, especially for those who are not willing or able to visit the much less accessible Ccolpa de Guacamayos area upriver; (2) a dual research station for national and foreign biologists; (3) a center of ethnobotanical research; (4) an educational center for tourists, and hopefully in the future, local student groups and teachers; and (5) a base from which activities such as timber extraction, gold mining, hunting, and many other types of economic activities can or could be monitored to some degree. It would be extremely unfortunate if this area were not somehow included in the Tambopata-Candamo conservation initiative. One solution would be to include the existing forests along both sides of the Tambopata from the mouth of the La Torre downriver to Infierno in some type of protected buffer zone that would extend east to the Pampas del Heath National Sanctuary, and then abut with the northeast boundary of the park somewhere to

ación humana en algunas de estas importantes áreas biológicas debería ser limitada o prohibida. Como un compromiso en último caso, la extracción forestal y de productos animales podría ser permitida a bajo nivel y cuidadosamente monitoreada, especialmente si estas actividades resultan en beneficio directo para las comunidades locales río abajo. En la región del alto Tambopata, vemos una clara y quizás única oportunidad de proteger una de las últimas secciones inhabitadas de ríos de aguas blancas en tierras bajas del este peruano. Con-cretamente, basados en ambos argumentos biológicos hechos anteriormente (y otras consideraciones discutidas más adelante), nosotros recomendamos la inclusión de aquella porción del río que queda entre la boca del Rio Távara y la boca del Río Malinoswski. Dicha acción nos parece crítica para un éxito a largo plazo más que cualquier otra iniciativa conservacionista en la región. El fallar en proteger los bosques a lo largo de esta porción del Tambopata, sin duda sería seguida por una colonización de esta área en los próximos 5 a 10 años, que a su vez podría resultar en un incremento de la presión sobre la llamada área núcleo del parque nacional al pie de montaña sobre los 400 m.

Un argumento adicional que puede fácilmente ser hecho para incluir el Tambopata sobre el Malinowski en el Parque deriva de nuestra sincera creencia de que la población local, especialmente aquella que vive cerca de la boca del Malinowski y río abajo por muchos kilómetros, podría últimamente beneficiarse más del desarrollo del ecoturismo a lo largo del alto Tambopata que de las presentes formas de agricultura o explotación maderera. La belleza escénica del área de la Ccolpa de Guacamayos, que atrae un inmenso número de guacamayos y otros loros, puede sin ninguna duda llegar a ser en América del Sur uno de los principales destinos para grupos de ecoturismo.

the southwest. Some type of compensation (or a new economic relationship between Explorer's Inn and the village) would necessarily have to be worked out with the Ese'eja.

Our biological surveys in forests above approximately 800 m on ridges bordering the Río Távara revealed the presence of lower montane vertebrate communities that differ markedly, in the case of birds and frogs, from those found in riverine and *terra firme* forests in the lowlands. Plant communities were less distinct within the elevational range covered on this expedition, but are known to also change markedly (in structure and species composition) at elevations above about 1500 m. This underscores the need for including all montane forest types within the National Park. Fortunately, most of the montane area proposed for protection is in steep, uninhabited terrain that will not likely attract colonists or economic exploitation in the near future. For this reason, it would be unfortunate to focus most of the effort to establish and maintain a protected area (national park) in the mountainous southwestern portion of the region, to the exclusion of the biologically distinct, more species-rich, and much more vulnerable lowland forests.

In summary, recent biological surveys along the upper Río Tambopata and Río Távara revealed the presence of a diverse array of forest types that differ markedly in terms of structure and species composition. For example, floodplains along the lowland portions of the Tambopata support several distinct types of forest (with exceptionally diverse floras and faunas) that should be included in any type of protected area established to maintain the biological diversity of the region. We confirmed the presence of additional, distinct montane plant and animal communities in the low mountains (500-900 m) that rise above the Río Távara, and know from previous work in nearby areas to the

Esta atracción, combinada con otros atractivos que ofrecen las instalaciones turísticas como el Explorer's Inn, Cuzco Amazónico, y el Manu Lodge, podrían y deberían ser el punto focal para revitalizar la una vez próspera industria en el sudeste peruano. Si esto ocurre, debemos solamente esperar que todos los involucrados en el negocio busquen caminos creativos y apropiados para involucrar a la población local en sus esfuerzos. Verdaderamente, una industria como ésta, que con el tiempo podría atraer miles de turistas, y podría resultar en una gran y más segura forma de beneficio económico que los presentes derivados del esfuerzo de la agricultura de subsistencia, y la extracción no-sostenida de madera, o la minería de oro.

Debería tomarse en fuerte consideración la inclusión en el Parque de ejemplos representativos de los llanos inundados y de los bosques de tierras altas en las terrazas aluviales altas a lo largo del bajo Tambopata en las cercanías del Explorer's Inn. Trabajando (en algún grado) con la población local durante 19 años, las operaciones de esta empresa turística han hecho posible la protección de mucho del bosque rivereño entre la propiedad, en la boca del Río La Torre, y la mas cercana comunidad de nativos Ese'eja en la parte mas baja del río, en Infierno. Aunque la relación entre los del Explorer's Inn y aquellos en la comunidad fué complicada y frecuentemente tirante, el resultado final de este continuo contacto ha sido la sobrevivencia de los bosques y la vida silvestre -y el potencial que ello representa para la investigación y el ecoturismo.

Contrastando con lo anterior, todos los bosques cercanos que bordean el Tambopata arriba del La Torre (todo el camino hacia el Malinowski), y río bajo hacia Puerto Maldonado, han sido reemplazados por agricultura y ranchos de ganadería.

... floodplains along the lowland portions of the Tambopata support several distinct types of forest ... that should be included in any type of protected area established ....

north and south that other distinct communities exist at higher elevations (e.g., 1500-2500+ m). We recommend that an attempt be made to include all of these ecosystems and communities in the proposed National Park. This would greatly increase the biological value of this park to both the nation and the world.

The ... preservation ... of the small Peruvian Pampas del Heath ... depends on the continued existence of the nearby Bolivian pampas ....

2 The Río Tambopata basin should be viewed as an integrated system, with all parts included in its management. The integrity of the upper watershed is important for the human communities below, not only for maintaining the quality of the river itself, but also as a production area for fish and game that are exploited downriver. A guardpost and educational center for the National Park could easily be established and maintained at the confluence of the Tambopata and Malinowski rivers.

Although the fauna of the lower river is more disturbed than that above, the Explorer's Inn Reserve, now completely flanked by colonists, has nonetheless managed to maintain good or increasing populations of the large mammals that were present at its inception (one monkey species was extinct, probably from overhunting, when the reserve was created). An important habitat type, large to huge blackwater lakes, is restricted to the lower Tambopata (a good example, Cocococha, is protected in the Reserve). Explorer's Inn, with its innovative resident naturalist program, has been a focal point for tourism and research that has set a standard for quality of both. The 15 years of research is a baseline for future longterm monitoring of environmental changes in the region. We recommend that the fully protected zone of the Explorer's Inn Reserve be extended eastward to join the Pampas del Heath National Sanctuary, to form a continuous protected corridor between the Ríos Tambopata and Heath. The administration of the Río Heath entity

Debido a la bien documentada importancia biológica, así como a su posición estratégica en la boca de La Torre (que fluye desde una gran área inhabitada juntándose al drenaje de la parte baja del Rio Heath), y por la larga historia de investigación ecológica en los bosques que los rodean, el Explorer's Inn merece una especial consideración como base de operaciones para cualquier tipo de unidad de conservación a establecerse en la región. Este ya sirve a numerosos propósitos entre los cuales se incluyen: (1) sus operaciones como albergue turístico, especialmente para aquellos quienes no están dispuestos o listos para visitar el área río arriba en la Ccolpa de Guacamayos; (2) una doble estación científica para biólogos nacionales y extranjeros; (3) un centro de investigación etnobotánica; (4) un centro de educación para turistas, para grupos de estudiantes y profesores locales; (5) un centro base desde el cual muchas actividades tales como la extracción maderera, la minería de oro, la caza, y muchas otras actividades económicas pueden o podrían ser monitoreados en algún grado. Sería extremadamente desafortunado si esta área no fuera de alguna manera incluída en las iniciativas de conservación del Tambopata-Candamo. Una solución podría ser incluir los bosques existentes a lo largo de ambos lados del Tambopata desde la boca de La Torre río abajo hacia Infierno en algún tipo de zona amortiguadora protegida que se extendería al este hasta el Santuario Nacional de las Pampas el Heath, y entonces unir con el límite noreste del parque con algún lugar al sudoeste. Algún tipo de compensación (o una nueva relación económica) entre el Explorer's Inn y la comunidad de los Ese'eja tendría que ser resuelta.

Nuestros estudios biológicos en los bosques sobre los 800 m. en la colinas que bordean el Río Távara revelaron la presencia de comunidades de vertebrados de bosque montano bajo que difieren marcadamente, en

could remain separate, and parts of the forest currently exploited for castaña could be zoned for managed use of that resource only, with no clearcutting or colonization.

3 Continued preservation of the larger neighboring Bolivian grasslands. The long term preservation of the biota of the small Peruvian Pampas del Heath, especially of its populations of large mammals, depends on the continued existence of the nearby Bolivian pampas (as already discussed by Hofmann et al. in 1976). We also hope that a binational agreement can be reached for protection and management of both banks of the river and all of the pampas.

4 Forest management practices that prevent excessive conversion of natural forest to bamboo in the region. The widespread presence of bamboo in Madre de Dios makes it necessary to approach with extreme caution all enterprises that involve clearing the forest. When forest is stripped for chacras, cattle, or other purposes, and then abandoned (as is almost universally the case), bamboo and not forest overtakes most of the disturbed areas. This can clearly be seen along the entire Río Tambopata, where abandoned communities dating from the rubber-boom era can now be easily identified as large bamboo patches.

Bamboo is a natural succession in disturbed areas in the region, but human activities have greatly increased its extent. Because it sprouts back up after burning, it is difficult to reuse the land in rotating swidden agriculture. Bamboo forests have an impoverished vertebrate fauna due to the drastic reduction of plant resource species. Because bamboo prevents the regeneration of forest for many decades or perhaps centuries, short-term logging or agriculture in this region carries a strong risk of long-term severe degradation of the high ground of its richest, floodplain forest

el caso de las aves y ranas, de aquellas encontradas en la riveras y en los bosques de tierra firme en las tierras bajas. Las comunidades de plantas fueron menos distinctivas dentro de este rango de elevación durante esta expedición, pero son conocidos los marcados cambios (en estructura y composición de especies) en elevaciones sobre los 1500 m. Esto subraya la necesidad de incluir todos los tipos de bosques montano dentro del Parque Nacional. Afortunadamente, la mayoría del área montana propuesta para protegerse está en las pendientes, en terrenos que no son habitados y que no parecen atraer a colonos o a una explotación económica en un futuro cercano. Por esta razón, sería desafortunado enfocar la mayoría de los esfuerzos en establecer y mantener una área protegida (parque nacional) en la porción montañosa sudoeste de la región, a la exclusión de los bosques de tierra baja biológicamente distintas, más ricas en especies, y mucho más vulnerables.

En resumen, estudios biológicos recientes a lo largo de la parte alta del Río Tambopata y Río Távara revelaron la presencia de diversos tipos de bosques que difieren marcadamente en términos de estructura y composición de especies. Por ejemplo, las llanos inundados a lo largo de porciones de tierras bajas del Tambopata presentan diferentes tipos de bosques peculiares (con excepcional diversidad de flora y fauna) que deberían ser incluídos en cualquier tipo de área protegida para mantener la diversidad biológica de la región. Hemos confirmado la presencia adicional de distintas comunidades de plantas y animales en las montañas bajas (500-900 m.) que se desarrollan sobre el Río Távara, y sabemos por trabajos previos que en áreas cercanas al norte y sur, de otras comunidades peculiares que existen a elevaciones más altas (por ejemplo, 1500-2500 + m). Recomendamos que

… las llanos inundados a lo largo de porciones de tierras bajas del Tambopata presentan diferentes tipos de bosques peculiares … que deberían ser incluídos en cualquier tipo de área protegida ….

ests where humans generally settle. We were disturbed to see that large strips of riverside forest just upstream of the Río La Torre had very recently been clearcut, but no crops planted, evidently in a effort by illegal colonists to gain title to unused land. Such practices could accelerate the bamboo takeover of the region without a single crop being taken from the land.

5 Management of the Pampas del Heath with deliberate burning. The health, diversity, and longevity of the Pampas del Heath depend on a program of deliberate and managed burns that will prevent the otherwise inevitable development of forest, thicket, and *Selaginella* from taking over the grassland. Such burning would secure the future of marsh deer and maned wolves in the pampas. To maintain the diversity of the habitat and the diversity of species it should not all be burned at once, nor every year. A rotating system of burning for different sections of the pampa—following the guidance (timing, wind-direction, etc.) of the people who have lived and hunted in the area for many years—must be implemented. It would be advisable to enlarge the area of open savanna by more intensive burning of those areas that were most recently savanna, such as the large isolated area of young regenerating pampa to the north of the current pampa.

6 Biological surveys and management plans for Pampas del Heath. Thus far, only one of the pampas has an access trail and has been partially surveyed. All other pampas need biological survey and management plans tailored to their individual vegetation conditions. In order to completely inventory the pampa mammals (or other vertebrates) and find out their distributions within the Sanctuary, a thorough inventory should be made of each isolated fragment. In any survey, a search

debe hacerse el intento de incluir todos estos ecosistemas y comunidades en el propuesto Parque Nacional. Esto incrementaría grandemente el valor biológico de este parque para el país y para el mundo.

2 La Cuenca del Río Tambopata debería ser visto como un sistema integrado, que incluya todas las partes para su manejo. La integridad de la vertiente alta es importante para las comunidades de río abajo, no solamente por el mantenimiento de la calidad del río, sino también como un área de producción de peces y animales para la caza que son explotados río abajo. Un puesto de vigilancia y un centro de educación para el Parque Nacional podrían ser fácilmente establecidos y mantenidos en la confluencia de los ríos Tambopata y Malinowski.

Aunque la fauna en la parte más baja del río está más perturbada que la de arriba, la Reserva Explorer's Inn, ahora completamente flanqueada por colonos, ha trabajado no obstante para mantener e incrementar poblaciones de mamíferos grandes que estaban presentes en sus inicios (una especie de mono estaba extinta, probablemente debido a la sobrecaza, cuando la reserva fue creada). Un tipo de habitat importante, los grandes lagos de aguas negras, están restringidos al bajo Tambopata (un buen ejemplo, Cocococha, es protegida en la Reserva). El Programa de residentes naturalistas del Explorer's Inn, ha sido un punto focal para el turismo e investigación que le ha dado calidad a ambos. Los 15 años de investigación son una base a largo plazo para futuros monitoreos de cambios ambientales en la región. Recomendamos que la zona total protegida de la Reserva Explorer's Inn sea extendida hacia el este para unirse al Santuario Nacional de las Pampas del Heath, formando un corredor protegido entre los Ríos Tambopata y Heath.

Figure 4 (upper right). Profile of floodplain forest on the lower Río Tambopata. The red flowering tree is an Erythrina *sp. (Leguminosae); the tall palms are* Iriartea deltoidea; *the short palms are* Astrocaryum chonta. *Photograph by Frans Lanting.*

Figure 5 (above). Río Távara, downstream from the Cañón del Távara. The high Távara ridges are in the background. Photograph by Frans Lanting. Figure 6 (lower right). Río Tambopata, showing its "braided" character upstream from its junction with Río Malinowski. The Cerros del Távara are in the background. Photograph by Frans Lanting.

Figure 7 (left). Three species of macaws (Blue-and-Yellow, Scarlet, and Red-and-Green) eating clay at the Ccolpa on the upper Río Tambopata. The behavior of macaws and other parrots at clay licks is still a mystery; presumably the clay contains essential salts and minerals, and it may also help to neutralize toxic chemicals found in the seeds eaten by these magnificent birds. Photograph by Andrew Kratter.

Figure 8 (right). Pampas del Heath. Pink-flowered shrubs are Macairea thyrsiflora (Melastomataceae). Photograph by Louise Emmons.

should be made for evidence of maned wolves (tracks are usually easy to find and identify). A census of the entire marsh deer population is needed, with particular attention to the numbers of young, so that the reproductive status of the population can be followed. This can only be adequately done by aerial survey (Schaller and Vasconcelos 1978). The census should be repeated every five years. At the same time, the status of the vegetation could be monitored.

7 The Heath Sanctuary has good potential for ecotourism. The pampas are beautiful, with stunning sunsets and sunrises, but they are also difficult to view. We suggest building a few simple fireproof "miradores," especially near wet zones frequented by marsh deer, and on high points. The natural salt licks (ccolpas) used by large mammals, especially the enormous one an hour upstream from Refugio Juliaca, would make excellent tourist attractions if they had mosquito-proof miradores. Some added protection against poachers may be needed however, because these licks are used by hunters. Another attraction could be a trail for adventure tourism linking Explorer's Inn and the Pampas del Heath.

8 Additional inventory research is recommended. Further inventory of small mammals at well-known research sites in Madre de Dios (e.g., Explorer's Inn Reserve) is needed so that we can reach the goal of knowing the entire mammal fauna of a locality in the upper Amazon basin (it is known for no species-rich site). In particular, different methods need to be used to collect the many insectivorous bats that should be present but are missing from the lists.

The largest vacuum in our knowledge of the regional fauna is of the elevational band from 1000 to 3000 m. Very little mammal or herpetological inventory has been done at

El manejo del Río Heath, como un conjunto, podría quedar separado, y partes del actual bosque de castaña explotado podría ser zonificado para el manejo de este recurso solamente, sin más tala o colonización.

3 La preservación continua de los pastizales vecinos bolivianos. La preservación a largo plazo de la biota de las pequeñas Pampas peruanas, especialmente de las poblaciones de mamíferos grandes, dependen de la continuidad de las cercanas pampas bolivianas (ya discutido por Hofmann et al. en 1976). También esperamos que el acuerdo binacional pueda lograr la protección y manejo de ambos lados del río y de toda las Pampas.

4 Prácticas de manejo del bosque que prevean una excesiva conversión de bosques naturales a bambú en la región. La extensa presencia de bambú en Madre de Dios hace necesario tomar extremas precauciones en todas las iniciativas de aclaramiento de bosques. Cuando el bosque es cortado para chacras, ganado, u otros propósitos, y después abandonado, (como es el caso generalizado), es el bambú y no el bosque quien retoma estas áreas. Esto puede claramente ser visto a lo largo de todo el Río Tambopata, donde después de la era del boom del caucho las comunidades abandonadas son ahora fáciles de identificar como grandes áreas de bambú.

El bambú viene como una forma natural de sucesión en áreas pertubadas en la región, pero la actividad humana ha incrementado sus extensiones. Debido a que estos brotan después de incendios, es muy difícil el reuso de la tierra con el método de agricultura rotacional. Los bosques de bambú tienen una fauna de vertebrada empobrecida, por la drástica reducción de especies de plantas. Dado que el bambú detiene la regeneración del bosque por muchas décadas o quizás siglos, la actividad maderera o agrícola

La preservación ... de las pequeñas Pampas peruanas ... dependen de la continuidad de las cercanas pampas bolivianas ....

that elevation south of the Río Madre de Dios. An expedition to inventory vertebrates intensively is needed in the upper Candamo part of the Reserved Zone.

9 Applied research on extractive resources and wildlife management should be encouraged. Research that would increase the local capacity to manage extractive resources such as castañas and other forest products would be useful. Developing research agendas in consultation with local communities could increase the value of research to the area and promote greater cooperation and full participation in management.

a corto plazo en esta región conllevaría un gran riesgo de degradación del suelo a largo plazo, en los bosques mas ricos, los de los llanos inundados, donde generalmente los humanos se asientan. Nos perturbamos al ver que grandes franjas de bosques rivereños justamente río arriba del Río La Torre habían sido recientemente cortados, pero no sembrados, evidentemente en un esfuerzo por parte de colonos ilegales de ganar títulos por tierras no usadas. Dichas prácticas pueden acelerar a que el bambú se apodere de la región sin que una sola cosecha sea tomada de la tierra.

5 El Manejo de la Pampas del Heath con quema deliberada. La salud, diversidad y longevidad de las Pampas del Heath dependen de un programa de quemas deliberadas y manejadas que preveerán el de otra forma inevitable desarrollo del bosque, matorrales, y *Selaginella* que tomaria sobre los pastizales. Dichas quemas asegurarían el futuro del ciervo de los pantanos y del lobo de crín en las pampas. Para mantener la diversidad del habitat y la diversidad de las especies, las pampas no deberían ser quemadas toda a la vez, ni cada año. Un sistema de rotación de quemas para las diferentes secciones de las pampas - siguiendo la guía (tiempo, dirección del viento, etc.) de la gente que ha vivido y cazado en el área por muchos años- debe ser implementado. Sería recomendable agrandar el área de savana abierta, con quemas mas intensas de aquellas áreas que recientemente fueron savanas, tales como las grandes áreas aisladas recientemente regeneradas al norte de la actual pampa.

6 Estudios biológicos y planes de manejo para las Pampas del Heath. Hasta ahora, solamente una de las pampas tiene una trocha de acceso, la cual ha sido parcialmente estudiada. Todas las otras pampas necesitan estudios biológicos y planes de manejo diseñados para sus condiciones individuales vegetacionales. Para inventariar completamente los mamíferos de las pampas (y otros vertebrados) y determinar su distribución dentro del Santuario, un cuidadoso inventario de cada fragmento aislado debería ser hecho. En cualquier estudio, una búsqueda de alguna evidencia del lobo de crín debe ser realizada (las huellas son usualmente fáciles de encontrar e identificar). Es necesario un censo de la población entera del ciervo de los pantanos, con particular atención en el número de jóvenes, de esta manera el estado reproductivo de la población puede ser determinado. Esto solamente puede ser adecuadamente hecho vía un estudio aéreo (Schaller y Vasconcelos, 1978). El censo debería ser repetido cada cinco años. Al mismo tiempo, podría monitorearse el estado de la vegetación.

7 El Santuario del Heath como un potencial para el ecoturismo.
Las pampas son hermosas, con impresionantes salidas y puestas de sol, pero ellas no son fáciles de ver. Sugerimos construir algunos simples "miradores" a prueba de fuego, especialmente cerca a las zonas húmedas frecuentadas por los ciervos de los pantanos, y en puntos altos. Los salitres naturales (ccolpas) usadas por mamíferos grandes, especialmente la ccolpa enorme a una hora río arriba del Refugio Juliaca, podría ser una excelente atracción turística si contaran con miradores a prueba de zancudos. Sin embargo, alguna protección adicional contra cazadores furtivos podría ser necesaria, porque estos salitres son tambien usados por ellos. Otra atracción pueden ser las trochas para turismo de aventura uniendo el Explorer's Inn y las Pampas del Heath.

8 Inventarios adicionales son recomendables.
Inventarios adicionales de pequeños mamíferos en los sitios bien conocidos para investigación en Madre de Dios (por ejemplo, Reserva Explorer's Inn) son necesarios para poder alcanzar la meta de conocer completamente la fauna de mamíferos de una localidad en la cuenca alta del Amazonas (los mamíferos pequeños son conocidos en muy pocas localidades ricas en especies). En particular, se necesitan usar otros métodos para colectar los muchos murciélagos insectívoros que debiendo estar presentes, se encuentran ausentes de las listas.

El gran vacío en nuestros conocimientos de la fauna regional está en las bandas elevadas desde 1000 a 3000 m. Muy pocos inventarios herpetológicos o de mamíferos han sido hechos en esas elevaciones al sur del Río Madre de Dios. Una expedición para inventariar intensivamente a los vertebrados es necesaria en la parte alta de la Zona Reservada Tambopata-Candamo.

9 Investigación Aplicada en Extracción de Recursos y manejo de vida silvestre debe ser alentada.
Investigación aplicada que aumentaría la capacidad local para el manejo de recursos extractivos, como castañas y otros productos sostenibles del bosque, sería útil. El desarollo de programas de investigación hecho en consulta con las comunidades locales, aumentará el valor de la investigación para el área, y promoverá una mayor cooperación y participación en su manejo.

Technical Report

METHODS

The 1992 rapid assessment in the Tambopata-Candamo Reserved Zone consisted of four separate trips. The first two were of Ortega, Ascorra, and Icochea to the Río Heath from 28 May to 8 June; and Gentry, Reynel, Nuñez, and Ortiz to the Cerros del Távara from 15 to 29 May . The second two trips consisted of Parker, Foster, Emmons, Rodríguez, Romo, and Wust to the Ccolpa de Guacamayos and Cerros del Távara from 22 May to 11 June; and Lamas, Foster, Emmons, Romo, Alban Castillo, and Wust to the Río Heath from 12-20 June. In addition, data is included here from previous research by several of the investigators, mainly at Explorer's Inn.

We were unable to get an aircraft to make overflights of this region, although we did have a quick look at the Bolivian side of the Río Heath in an overflight made on another expedition. However, we were greatly assisted by a color composite Landsat, Thematic Mapper satellite image (TM bands 3,4,5) of the Tambopata-Madidi region. This image allowed us to pinpoint certain vegetation types for field examination, especially on the Río Heath, and to make general observations regarding the extent of the different vegetation types described in this report.

Gentry, Reynel, Nuñez, and Ortiz collected plants and data on all woody species found in a series of transects 50 m long and 2 m wide (totaling 0.1 ha) at two sites in the Cerros del Távara. Prior to this trip, Gentry did another transect series at one site on the Río Heath and inventoried several one-hectare plots at Explorer's Inn on the Río Tambopata. Foster made ground surveys of as many habitats as possible at each site except for Explorer's Inn. This included qualitative assessment of plant community composition and structure at each site. Alban Castillo and Foster made collections of the plants on the Pampas del Heath.

Parker surveyed and documented the presence of bird species with the aid of tape recorders and unidirectional microphones (Parker 1991). On this trip he surveyed the birds of the Távara and Ccolpa de Guacamayos for the first time, and on a previous trip he surveyed the birds in the Heath Sanctuary (Appendices 1-4). Parker, Donahue, Schulenberg, and others had surveyed the birds at Explorer's Inn on an annual basis since 1976. Voucher specimens were collected for a few unusual or difficult-to-identify species (see Parker 1982).

Mammals were censused by day and night observation along trails (about 35 h at the Ccolpa de Guacamayos, and 32 h at Cerros del Távara), by trapping (672 trap/nights), and by mist-netting (55 net-nights) (Appendix 5). Monica Romo was chiefly responsible for collecting bats on the expedition and Louise Emmons for non-flying mammals. Emmons made daily and nightly walks in the forest and savanna habitats studied, recording all species seen, heard, or identified by their tracks. Previous records from the same localities by Walter Wust (1989 and pers. comm.) were a significant contribution to the list. The Explorer's Inn list is based on prior work in the area (Appendix 6).

Mammal surveys were done by two groups that visited the Pampas del Heath Sanctuary (SNPH) separately. The first group composed of Ascorra and colleagues visited the SNPH from 28 May to 8 June 1992; while the second group, including Romo and Emmons, worked in the SNPH from 12-20 June. Ascorra collected mammals in both the forest and pampa, while Emmons and Romo concentrated all collecting in the open pampa, its edge, and gallery forest inside the pampa, but also made observations in the forest. Romo chiefly collected bats, Emmons non-flying mammals, and Ascorra both. Mammals were identified during day and night observation walks; by calls, tracks, bones or feces; by trapping with Sherman, Tomahawk, and Victor rat traps; and by mist netting both near the ground and at 10-15 m in the subcanopy (Ascorra). Other mammal species were listed by residents of the area (Appendix 7).

Rodríguez collected the herpetofauna in the Cerros del Távara and near the Ccolpa. Her observations included visual encounters and recognition of frog calls (Appendix 8). Icochea collected on the Río Heath, where he searched bodies of water and trails through the forest and pampa both during the day and at night. A few frogs and a turtle were caught in a fish net by Ortega. Large reptiles were observed and recorded, but not collected (Appendix 9).

Ortega collected fish on the Río Heath. Specimens were collected with 4 x 1.5 and 6 x 1.8 m seines with 4 and 6 mm mesh respectively; cast nets and a 1.5 m dip net were also used, with which intensive sampling (10 repetitions) was carried out using the net appropriate for each collecting site (Appendices 10 and 11). Each sampling site was photographed in addition to the habitat notes taken. Species identifications were made based primarily on keys, recent revisions, original descriptions, and comparative material in the ichthyology collection at the Museo de Historia Natural in Lima. The fish classification used follows that proposed by Greenwood et al. (1966), with modifications adopted by Nelson (1984), Ortega and Vari (1986), and Ortega (1991).

Lamas collected and recorded butterflies in the Pampas del Heath for the first time, and contributed his species list from many years of work at Explorer's Inn (Tambopata). Butterfly species at both sites (Tambopata and Pampas del Heath) were recorded either visually or by collecting them by conventional methods (using entomological nets and various baits) (Appendices 12 and 13). At Tambopata, only unusual-looking or hard-to-identify species were collected, as the area is quite well surveyed after 13 years of effort. At Pampas del Heath, collections were made in three different habitats: forest, savanna, and gallery forest within the savanna. No collections were made at the Heath riverbanks, but the species observed in that habitat were recorded. Almost equivalent time and effort were dedicated to both savanna and gallery forest, and less was spent on the forest habitat surrounding the savanna, because its fauna was similar to that of Tambopata.

Vouchers of plant and animal collec-

Figure 3. Map of the
Tambopata-Candamo
Reserved Zone with
major study sites
indicated.

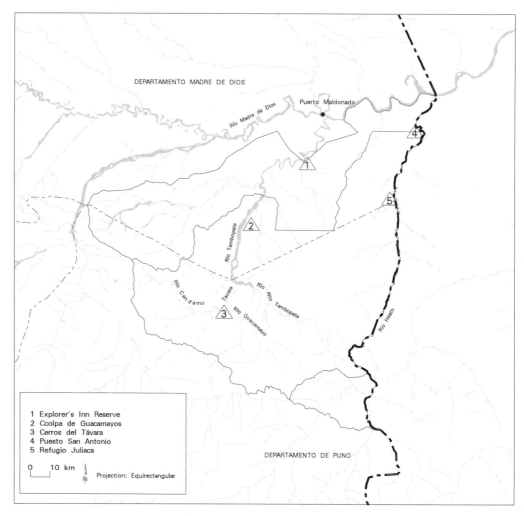

DEPARTAMENTO MADRE DE DIOS

Puerto Maldonado

Rio Madre de Dios

Rio Tambopata

Rio Candamo

Tavara

Rio Guacamayo

Rio Alto Tambopata

Rio Heath

1 Explorer's Inn Reserve
2 Ccolpa de Guacamayos
3 Cerros del Távara
4 Puesto San Antonio
5 Refugio Juliaca

0   10 km

Projection: Equirectangular

DEPARTAMENTO DE PUNO

tions are deposited in the Museo de Historia Natural of the Universidad Nacional Mayor de San Marcos in Lima (MUSM). Duplicate plant collections are in part at the Missouri Botanical Garden (St. Louis, USA) and Field Museum of Natural History (Chicago, USA). Copies of bird recordings will be housed in the Library of Natural Sounds, Cornell University (Ithaca, USA). Duplicate mammal, amphibian, and reptile specimens are deposited in the National Museum of Natural History, Washington, DC (USNM). For further information on methods see Gentry (1982), Parker and Bailey (1991), and Parker (1991).

Conclusions regarding the relative conservation importance of the areas visited in the present study are based on comparisons of data compiled at numerous sites throughout the Neotropics by us and many other researchers. Criteria used for making judgments regarding biological priorities include the total biodiversity of an area, levels of endemism (percentage of species and genera restricted to relatively small geographic areas), presence of rare or threatened forms, and degree of risk of extinction of species and/or communities at a national or global scale.

OVERVIEW OF THE LANDSCAPE AND VEGETATION OF THE TAMBOPATA-HEATH REGION OF PERÚ (R. FOSTER)

The Peruvian Andes drain toward the Atlantic Ocean in two widely diverging watersheds:

CONSERVATION INTERNATIONAL   **Rapid Assessment Program** one toward the north (toward the Equator) joining the central Amazon River headwaters, and the other toward the south to join the Río Madeira, the main southwestern tributary of the Amazon. These two watersheds are separated at the base of the Andes by the Fitzcarrald Divide—the Urubamba and Ucayali rivers to the north, and the Manu and Madre de Dios rivers to the south.

In southeastern Perú, after the Manu and Alto Madre join to form the **Río Madre de Dios**, this river continues southeast, gradually distancing itself from the Andes, but still picking up tributaries out of the Andes to the south. In order, these principal tributaries are the Inambari, Tambopata, Heath, and finally the Beni as the river swings north into the Río Madeira.

The **Río Tambopata** and the Madre de Dios join at Puerto Maldonado after passing through a region of high, ancient terraces composed of a sandy clay. These flat-topped but dissected terraces are the westernmost extension of a band that stretches through eastern Madre de Dios and Pando, Bolivia, into central Brazil along the southern "slopes" of the Amazon basin. Visitors to Explorer's Inn on the lower Tambopata are familiar with this high terrace found on the back part of the area near Cococcocha.

The **Río Heath**, which joins the Madre de Dios where it passes into Bolivia and whose length is part of the border between Perú and Bolivia, has a catchment basin much smaller than that of the Tambopata and its major tributaries. It drains only the lowest ridges of the Andes. Its tributaries do not greatly expand its breadth. Possibly the Río Tambopata itself used to flow out of the Andes directly into what is now the Heath basin before the slight uplift which separates them diverted it to the northwest. Just above the floodplain along the Heath are extensive low terraces, called *sartenjales*, with a distinctive thick vegetation. This landform and vegetation is missing or only rarely found along rivers north of the Heath.

The floodplain of the Tambopata and its tributaries has the usual array of different successional bands and different ages of terraces associated with differences in the frequency of flooding. The composition of the forest changes gradually with the age of the terrace. Oxbow lakes formed from cut-off river meanders are not common in this area, but where they occur they contribute importantly to the diversity of the flora and fauna.

Between the Ríos Tambopata and Heath, the high, ancient terraces stretch back toward the mountains, grading almost imperceptibly into the more recent alluvium fanning out from the foothills. Similarly, the narrow floodplains of the small rivers and streams that fissure these terraces gradually merge with the alluvial plains of their headwaters where "terrace" and "floodplain" lose their meaning. This large transition zone, which is even more extensive in Bolivia, has yet to be explored by biologists, and the vegetation can only be described as fairly low, open forest with high densities of palms.

On the terraces, the sandy soils on the lower part near the Madre de Dios and Tambopata give way to a heavy impermeable clay on the higher part. These two substrates have dramatically different vegetation: the former with a well-drained, diverse forest abundant in *castaña* (brazil nut); the latter with a poorly drained forest regenerating from what was an open grassy savanna—or the still existing savanna itself, i.e. the Pampas del Heath. Within these terrace types are numerous variants usually reflecting the degree of drainage and of temporary inundation.

In vegetation, the low foothill ridges apparently have much in common with the sandy terraces, due both to more drainage and the frequency of landslides on their mar-

The Río Tambopata and the Madre de Dios join at Puerto Maldonado after passing through a region of high, ancient terraces composed of a sandy clay.

gins. Both factors apparently are sources for the natural maintenance of bamboo stands. Large stands of bamboo can be spotted throughout the foothill region of this area. If our interpretation is correct, most of these are natural stands. The bamboo are maintained in a natural reservoir on the small landslides, and expand out into larger disturbances that happen nearby such as washouts, massive landslides, or major windthrows. However, colonization can only occur when the disturbance coincides with the very infrequent fruiting events of the bamboo species.

Along the rivers, except for the small landslides of the high terraces and an occasional large colonization surface formed when the river changes directions, the bamboo stands seem to be mostly from human disturbance in the form of large clearings. The bamboo colonization of these clearings will only occur if the forest cutting and abandonment coincides with the fruiting of the bamboo and the close proximity of a seed source.

The mountain vegetation is in its lower part similar to that of the foothills except that it is less prone to drying out. The major vegetation differences in the mountains are the result of different chemistry in the rock substrate exposed and different probability of exposure to periods of drought—especially in the ever-wet areas with frequent cloud contact.

The differences in vegetation and physiography summarized here are described in greater detail for the different localities visited. Only some of the important differences in vegetation can be seen from the satellite image presented here (Frontispiece). These are the vegetation types that have both large differences in the way they reflect light and are of sufficient contiguous size that they show up in contrast to their surroundings. From reading the individual locality accounts, however, it will be clear that much of the important diversity in habitats and plant species composition is not easily mappable using these images. Perhaps in the future the quality and cost of satellite images will improve to the point where we can map much more of the important habitat diversity. In the meantime the images are a useful tool to map some of the habitats and focus attention on the areas of our ignorance. What will never change is the need for more ground surveys to assess the composition and understand the dynamics of these habitats.

Detailed plant lists are not included in the appendices because determinations are still proceeding. The plants of Tambopata-Candamo will be published separately.

RÍO TAMBOPATA

The Landscape (R. Foster)

Moving upstream from the mouth until it divides, the Río Tambopata follows a jerky course characterized by many long straight stretches and few meanders. Its course is largely confined and stabilized by the high terraces on either side. There are many other levels of lower, recent terraces along the Tambopata. The Explorer's Inn buildings are on one of the highest of these recent terraces, a habitat that seems to be rare in the Tambopata drainage.

Two-thirds of the way to the mountains, heading upriver, the river splits and changes completely. The right fork, **Río Malinowski**, flows slowly through a broad recent floodplain where it has been actively meandering following a variety of tracks, and then up to the lower Andes. It is now eroding the relatively narrow band of high old terrace that separates it from the Madre de Dios. The left branch of the Tambopata becomes a fast, shallow, braided river, changing channels from side to side while depositing the alluvium coming straight down from the Andes. These two river branches are analagous (but on a smaller scale) to the braided Alto Madre de Dios that joins the meandering Manu—one comes steeply, straight out of the mountains and joins the other, which comes gradually from the side out of an alluvial basin.

On abruptly reaching the mountains, the Río Tambopata splits in two with one branch, the **Río Távara** going straight ahead into the mountains, the other, the **Río Alto Tambopata** swinging south and east along the base of the Andes before finally making penetration close to the headwaters of the Río Heath. The Alto Tambopata drains the high reaches of the Andes, including dry forest valleys. The Távara only drains the wet lower ridge systems.

The Távara, in its short course between the Tambopata and its division into the **Ríos Candamo** and **Guacamayo**, splits two large ridges and a couple of small ones. The large ridges reaching 1000 m, here called the Cerros del Távara, are themselves two sides of a single chain that splits apart temporarily in this region, creating two small east-west draining valleys on either side of the central Távara. Entering this split between ridges on the north side, the Távara passes through a narrow, steep-sloped canyon a couple of kilometers long. The passage through the southern ridge is less dramatic.

The **Candamo** is a relatively peaceful river on its lower reaches, interrupted by occasional tumultuous rough stretches, but the ridges being so close on all sides give it precious little room to meander. The **Guacamayo** is a wild, rambunctious river from the start, with a rapidly shifting narrow channel full of large boulders and frequent landslides. This either reflects a different geological structure or a difference in the rate the Andes are rising in this area.

Cerros del Távara (R. Foster)

The lower ridges cut by the Río Távara are the characteristic red clay rock of Tertiary origin found throughout the length of at least the Southern Andes as they continue to grow out into the Amazon Basin. This rock usually produces soils fairly rich in nutrients and higher in cation exchange capacity compared to that from the older rock sediments below. But it is certainly not as rich as the recent alluvial floodplains. The older rock sediments are exposed on the high ridges of the Cerros del Távara and in the canyons that pass through them. The most resistant to weathering seem to be quartzites that give rise to a vegetation much more characteristic of acidic or low nutrient availability. These uplifted strata are in close alternation with other strata less extreme in their properties and the result is sometimes a mix. One of the clearest places to see the distinction is on the north slope of the northeast ridge where a series of layers are exposed. One hikes up an A-shaped shield based on quartzite to a crest, then one drops down a little onto some softer, richer material; then one starts upslope again on another quartzite shield.

The advance team of Gentry, Reynel, Nuñez, Ortiz, and crew sampled the Cerros del Távara forest with transects at each of two sites: (1) a lower site at 400 m on the slope of a low ridge two bends upriver from the mouth of the Távara where it joins the Alto Tambopata; and (2) a higher site at 700-800 m on the upper part of the SE Cerro del Távara ridge reached from the Río Guacamayo.

The main team did not find the lower site of the advance team, but it did stop for a few hours to climb Cerro Mirador, the foothill ridge at the mouth of the Távara. Because of flooding, a forced camp was made for one night in the middle of the Távara canyon. The main basecamp was at 330 m on the east bank

The lower ridges cut by the Río Távara are the characteristic red clay rock of Tertiary origin found throughout ... the Southern Andes

of the Távara—right at the junction of its two large tributaries, the Guacamayo and the Candamo. This camp was situated on a high clay levee, 20 meters from the steep slope of the SE Cerro del Távara. Our trail of several km went straight up the slope to about 600 m, then followed the undulating ridge to increasingly higher elevations, reaching 900+ m.

This trail met, at about 700 m, the other access trail to the same ridge made by the advance team going up to the northeast from a narrow rocky floodplain on the Río Guacamayo. This latter trail is more heterogeneous, having some exposure of quartzite boulders, crossing streams, and passing through a large jumble of debris and vegetation tangle at the bottom of an old landslide. A third trail was made from downriver below the Cañón del Távara, going up the NE Cerro del Távara ridge from 300 m to 950 m, and on a different substrate.

Vegetation of Cerros del Távara (R. Foster)

Substrate Differences

The vegetation reflects the geological differences, with the greatest contrast in flora between the quartzite-based, sandy clay soils and the non-sandy soils. The differences in the forests of the NE & SE Távara ridge are compared in Table 1. In addition, the sandy soil flora differed from the clay soil flora in that it had a greater density of straight stems 10-30 cm in diameter, a more open interior, less disturbance, and fewer lianas.

Altitudinal Differences on the SE Távara Ridge

The forest on the slopes has emergent trees about 35-40 m tall; the rest of the canopy is about 30-35 m tall. This continues to about 800 m altitude, above which most of the trees are smaller, and eventually some "perpetu-

ally disturbed" low forest is reached above 900 m on ridge crests.

The conditions throughout the gradient are highly humid and may be slightly wetter near the river. It qualifies as wet lowland forest, but is not the superhumid lowland cloud forest found in some lower Andean localities (e.g., the base of the Yanachaga-Chemillén).

Like most wet ridges, this one has a very mixed forest with a considerable difference in composition from hectare to hectare. The plant communities at the base of the ridge and at 800 m are not significantly different. Many of the species on the high rocky floodplain are also found at the highest elevations. Most conspicuous among these species is *Dipteryx micrantha*, which grows at intervals as a large emergent all the way up the ridgecrest into the most moss laden forest at 825 m. *Dipteryx* is more characteristic of lowland, floodplain forest.

The differences one finds seem to have more to do with the regeneration of landslides, which are not especially frequent on these large ridges. The deposits of debris at the bottom of the ridges are often of a soupy consistency, very unstable, full of tangles, bamboo, etc. Gaps are relatively infrequent, but there were several areas of recent lightning strikes that killed several trees at once.

On the ridge crest, after it flattens out above into a series of undulations, each progressively higher from 550 to 800 m, the depressions or notches in the ridge are frequently foggy and dripping with moss almost characteristic of cloud forest. Above 800 m, the forest is more constantly mossy, but still there are very few species (Table 2) that could be considered representing a new community—at least one cannot distinguish unfamiliar species from the usual background noise of novel species one encounters walking through forest of the same habitat. To the contrary, many of the new species one encounters are typical

Table 1. Differences in the forests of the NE and SE Távara ridges. Despite these differences, many plants are found in common between the two soils and canopy heights are about the same.

Quartzite-based, sandy clay soils (NE)	"Regular" clay soils (SE)
Brosimum utile dominant	*Brosimum utile* rare
Geonoma deversa	*Geonoma arundinacea*
Rinorea guianensis absent	*Rinorea guianensis* present
Metaxya fern superabundant	*Metaxya* fern rare
Chrysobalanaceae frequent	rare
Vochysiaceae frequent	rare
Melastomataceae different	Melastomataceae diff.
Rubiaceae fewer small spp.	Rubiaceae more small spp.
Cybianthus frequent	rare
treeferns different	treeferns different
Psychotria erecta common	absent
Tococa guianensis common	absent
Aparisthmium cordatum common	absent
Pourouma mollis frequent	rare
Marantaceae different	Marantaceae different
Aspidosperma (convoluted sp.) common	rare
Heteropteris common	rare
absent	*Dipteryx* common
Trichomanes elegans common	absent
Anaxagorea different	*Anaxagorea* different
Myrtaceae frequent, different	Myrtaceae diff.
Hernandia rare	*Hernandia* common
Menispermaceae (fuzzy) common	absent

Table 2. Possible cloud forest species on the SE Távara ridge top.

Philodendron (small leaf, giant wing petiole)

Hyospathe (tiny leaf and inflorescence)

Myrcia

Hydrangea sp.

Psychotria sp.

Ladenbergia sp.

Siparuna sp.

Guzmania sp.

(Ericaceae) sp.

Clematis sp.

lowland species and the surprise is to find them so high.

More than 80 orchid species were collected in this area by Percy Nuñez during the visit by the advance team. Many of these are probably also found at higher altitudes in cloud forest.

On the east slopes of the Andes, cloud forest conditions are usually approached near 1000 m and are in full force by 1,500 to 2,000 meters. Such conditions can be approximated at much lower elevations if a ridge is isolated from the main mountains, if clouds come in low off the ocean, in notches or ends of principal ridge systems where clouds "slip through," in areas where clouds are "trapped" in a basin, or where there is considerable "clear night condensation." These conditions need to be met with little interruption for most of the year. The tops of the Távara ridges can only marginally be considered cloud forest.

Most Common and Characteristic Plants

Emergent Trees: *Cedrelinga, Buchenavia, Cedrela, Dipteryx, Sloanea* [rare]. On both north and south ridges at 700-800 m we encountered a few emergent *Podocarpus* sp. They have straight trunks and stem diameters between 50 and 100 cm. The adults were in or on the edges of old landslides, but many young seedlings and a few juveniles were found throughout the understory. The *Cedrelinga* and *Cedrela* are mostly on the lower slopes, but some were found on landslip or blowdown colonization above 800 m. They are frequently a meter or more in diameter.

Canopy Trees: *Hevea, Hernandia, Calophyllum, Qualea, Inga, Iriartea deltoidea, Tapirira, Huertea, Cecropia sciadophylla, Pourouma guianensis, Pouteria*, Lauraceae, *Matisia cordata, Hymenaea oblongifolia, Sloanea fragrans, Qualea, Hyeronima, Senefeldera, Nealchornea, Otoba* [orange],

Otoba parvifolia, Minquartia, Rinorea guianensis, Eschweilera. There are very few *Ficus*—perhaps the frequency of rain prevents pollination. There are two species present of *Tachigali*, the "suicidal" tree. In one of these species (trunks up to 80 cm diameter), many had just finished fruiting and were obviously dying.

Understory and Gap Trees: *Pausandra* sp. (Euphorbiaceae) is superabundant. Other common genera and species are *Protium, Ladenbergia, Miconia, Naucleopsis, Perebea, Pseudolmedia, Micropholis, Virola, Wettinia, Socratea salazarii, Quararibea ochrocalyx, Diospyros*, Myrtaceae, *Quiina, Guarea, Coccoloba*.

Shrubs: *Geonoma arundinacea, Carpotroche, Psychotria* (*Cephaelis*) sp., *Faramea* sp., *Stylogyne ramiflora, Chrysochlamys ulei, Miconia paleacea, Miconia* spp., *Cyathea* spp., *Pithecellobium macrophyllum, Pentagonia* cf. *macrophylla, Mollinedia, Anaxagorea, Rinorea, Esenbeckia, Galipea, Neea*.

Herbs: *Maieta, Tococa, Clidemia, Tectaria, Ischnosiphon, Monotagma, Renealmia, Costus, Dieffenbachia humilis, Spathiphyllum, Besleria, Danaea, Geophila*.

Trunk Epiphytes: *Philodendron acreanum, Polybotrya, Philodendron ernestii, Microgramma fusca, Lomariopsis japurensis, Drymonia*.

Lianas: *Byttneria, Hydrangea, Tetracera, Bauhinia*, Bignoniaceae, *Machaerium, Coccoloba, Strychnos*.

Other Components of the Vegetation

Species Occurring in Discrete Clumps (nonclonal): The following plants have seeds dispersed by an explosive mechanism which apparently is the cause of their being strongly clustered. This is often so extreme that the species may be the most abundant plant in one area (e.g., 1 ha) and completely absent in an adjacent area of the same habitat.

On the east slopes of the Andes, cloud forest conditions are usually approached near 1000 m and are in full force by 1,500 to 2,000 meters.

Annonaceae

Anaxagorea sp. 1
Anaxagorea sp. 2
Anaxagorea sp. 3

Euphorbiaceae

Acidoton venezuelensis
Euphorbia sp.
Mabea maynensis
Mabea sp.
Pausandra sp. - dangling frt.

Rutaceae

Galipea tubiflora?/jasminiflora?
Metrodorea sp.
Esenbeckia amazonica

Violaceae

Rinorea guianensis
Rinorea viridifolia
Rinorea lindeniana
Rinorea sp.

Palms: *Iriartea deltoidea* is common throughout the slope and is the most abundant large palm. *Oenocarpus bataua*, *Astrocaryum "macrocalyx," Oenocarpus mapora*, *Socratea exorrhiza*, and *Euterpe precatoria* are occasional to rare. The *Euterpe* is sometimes in aggregations (unconnected) of 6-12 stems. Of the medium-sized palms, *Socratea salazarii* and *Wettinia* sp. are common, the latter more frequent higher on the ridge. Small palms include a superabundance of *Geonoma arundinacea* on most ridge slopes, and *Geonoma deversa* on the sandy soil slopes. Other small palms such as *Bactris* cf. *actinoneura*, *Bactris simplicifrons*, *Bactris* (acaul), *Geonoma brongniartii*, *Geonoma* (thick stem), *Chamaedorea leonis*, *Chamaedorea* sp., *Hyospathe elegans*, and *Hyospathe* sp. are either occasional or rare. The climber *Desmoncus* is occasional throughout. The ridges are apparently not high enough to support *Dictyocaryum*, and *Attalea* s.l. is apparently absent.

Bamboo: On disturbed sites along the river, bamboo, "paca" (all apparently *Guadua weberbaueri*), is seen in small clumps. On major landslides of the lower slopes there are a few large stands of almost solid regenerating bamboo. Otherwise there is very little in the Távara area.

Chacras del Diablo: Within the forest understory, usually associated with old, small landslides and *Atta* ("leafcutter") ant colonies, are open areas with a distinctive group of species. In parts of Perú these are referred to as "chacras del diablo"—the devil's farms. The herb, shrub and small tree species present are a mixture of plants with special structures inhabited by small ants (e.g., *Tococa caquetense*, *Maieta guianensis*, *Clidemia* sp., *Pleurothyrium krukovii*, etc.), and others without ants such as a *Clidemia* with dense pink hairs, Gesneriaceae, and numerous ferns.

It is hypothesized that the landslides are colonized by *Atta* colonies that take advantage of the fast growing regeneration vegetation. These areas in turn can only be invaded by understory species that are avoided by the *Atta* ants.

Other Sites in the Távara area

Cañón del Távara

Halfway up the Río Távara, the river passes through a large ridge which channels the water to about 75 m width between large abutments of grey rock and steep slopes. During flood, the river rises about 5-6 m, scouring the rock sides of all but the most resistant plants. This distinct community is characterized by plants with very narrow leaves and flexible stems. The largest of these is *Pithecellobium longifolium*. These are mostly confined to the high water mark where they extend out over the river as they do on virtually every fixed bank river in the upper Amazon.

It is hypothesized that the landslides are colonized by Atta colonies that take advantage of the fast growing regeneration vegetation.

Other woody species include three species of Myrtaceae (*Eugenia riparia*, *Myrcia* [2 spp.]), *Erythroxylum*, *Acalypha*, *Miconia riparia*, *Cestrum*, *Ardisia huallagae*, and occasional *Calliandra*. Herbs on the rocks include *Koellikeria erinoides*, *Cuphea*, and *Sauvagesia*.

As the valley opens out on either end of the canyon, the zone is reduced to 2-3 m, or missing altogether when the banks are of Tertiary red clay or recent gravel alluvium. The beaches are mostly of rounded rocks and pebbles and if relatively stable, get covered with *Calliandra angustifolia*. Common festooning lianas are *Combretum fruticosum*, *Byttneria pescapraefolia*, *Mucuna rostrata*, and *Cratylia argentea*.

Cerro Mirador

Near the mouth of the Río Távara is a low ridge with a trail going up the south side for a view out over the lowlands and the abandoned village site of Astillero. The south slope and ridgecrest of Cerro Mirador appear to be Tertiary rock, but fairly poor soil intermediate to that of the two Távara ridges. It is fairly disturbed in the lower part, but the upper ridge has open understory. The forest is dominated by *Senefeldera*, *Hevea*, *Rinorea guianensis*, *Cedrelinga*, *Tachigali*, a few large *Dipteryx*, and it has more *Ficus* than the Távara ridges. There is an even mixture of palms. This ridge is perhaps more species rich in trees and shrubs as a whole than the main Távara ridges, but has more dominance by a few species.

Floristics and Phytogeography of Cerros del Távara (A. Gentry)

Each sample (0.1 ha) is the sum of 10 transects, each 2 m wide by 50 m long. The transects are usually contiguous but can differ in compass orientation. All woody stems greater than 2.5 cm diameter at 1.3 m height—including lianas, trunk climbers, and hemi-epiphytes—are included in the sample.

Lower Távara

While the lower Távara site is floristically very typical of lowland Amazonian forest, with dominance of Leguminosae trees (29 spp.) and Bignoniaceae lianas (12 spp.), it has unusually high diversity for southern Amazonia. The 187 species >2.5cm dbh in the 0.1 ha sample is significantly more than occur in the adjacent lowlands along the Ríos Madre de Dios and Tambopata (average: 149 spp.).

The most unusual floristic component of the sample as compared to others from Madre de Dios is the prevalence of Melastomataceae, with 13 species represented in the sample, perhaps some kind of world record for this family. While Melastomataceae are generally better represented in Madre de Dios forests than in those of the Iquitos area, equivalent samples at Cuzco Amazónico and Explorer's Inn have 0-7 melastome species, depending on substrate, less than half as many as in the lower Távara sample.

Melastomataceae are often considered indicators of disturbance and it is possible that the greater topographic relief in the steep foothill ridges of the Távara is correlated with a higher frequency of landslides or other disturbance. It is also possible that the unusual Melastomataceae representation is part of a generalized cloud forest effect associated with a moister climate at the base of the mountains, or possibly of ecologically more dynamic forests associated with greater productivity on richer soils. It should also be noted that this family is unusually prone to local endemics; therefore, even though most of the melastome species remain unidentified, it is likely that a significant fraction of them will prove to be conservationally important local endemics.

The rest of the important families in the lower Távara sample—Moraceae, Lauraceae, and Annonaceae—are exactly the same families that are prevalent in equivalent samples in most of Amazonia, including the lower Tambopata and Puerto Maldonado area lowlands. Euphorbiaceae, Rubiaceae, and Palmae are similarly well-represented both at Explorer's Inn and on the lower Távara.

In addition to Melastomataceae, families that are somewhat better represented on the lower Távara than at Explorer's Inn include Meliaceae (7 vs. 1-4, av. 2.5), and especially Sapindaceae (6 vs. 2) and Elaeocarpaceae (4-5 vs. 1). On the other hand, Myristicaceae, Chrysobalanaceae, and (perhaps) Sapotaceae are noteworthy for their low abundance in the lower Távara sample. The latter families are those that tend to be associated with poorer soils and their low representation is likely due to the relatively rich soils of the lower Távara sample site.

Although the familial composition of the lower Távara forest is similar to that of adjacent lowland Amazonia, the species composition is completely different. Conservationally, this is the most important aspect of the foothill forest in the Tambopata-Candamo area; it is a completely different set of taxa than are represented in the lowlands.

Higher Távara (provisional interpretation of data by R. Foster)

The sample of the SE Cerro del Távara ridgetop has a species total (ca. 232 species) that is considerably higher than either the lower Tambopata-Maldonado area or the lower Távara site. Sixty-one families are represented compared to the 51 at the lower site. In part, this reflects the greater stem density at the higher Távara site (444 individuals vs. 357), as well as the heterogeneity of the forest. The distribution of species among families is much more equable at the higher site, though Leguminosae and Moraceae still stand out with 19 and 12 spp. But the Melastomataceae,

Bignoniaceae, and Lauraceae are represented by less than half as many species at the higher site. Families with much better representation at the higher site are Myrtaceae, Violaceae, Guttiferae, Nyctaginaceae, and Rubiaceae.

Birds of the Cerros del Távara (T. Parker, T. Schulenberg and W. Wust)

The avifauna of this ridge is dominated by two elements, Amazonian species that are at or nearing their upper elevational limit, and species largely or entirely restricted to a narrow band of moist forest on the lowermost slopes of the Andes (Appendix 1). Due to this elevational restriction, and the high rate of forest clearance within this zone, this lower montane zone is one of the most critically threatened forest habitats in all of South America.

In the brief time spent at the Cerros, 41 bird species were recorded that are not present at the Ccolpa or the Explorer's Inn Reserve. However, three of these (*Tinamus tao*, *Myobius* [*barbatus*], *Tachyphonus rufiventer*) are known from Cocha Cashu. Almost all of the remaining 38 are species typical of or completely restricted to lower montane forests. Given that the Cerros del Távara is a relatively low ridge that would be at or near the lower elevational range of many of these species, this is an impressive figure.

A surprising find at the Távara was *Pipreola chlorolepidota*. This small, rare cotinga is not known from Bolivia, and indeed had not previously been recorded south of central Perú.

Highlights among the birds recorded at the Cerros are two species with particularly small and fragmented distributions, the Black Tinamou (*Tinamus osgoodi*) and the Horned Curassow (*Pauxi unicornis*). The tinamou has been recorded only from lower montane for-

... this lower montane zone is one of the most critically threatened forest habitats in all of South America.

est in Cuzco and Madre de Dios, Perú (with a highly disjunct population in central Colombia). The curassow is known only from a few sites in Bolivia and from the Cerros del Sira in central Perú. Apparently it is even more restricted than are most species of the lower montane zone, as it belongs to a species assemblage that is confined to ridges that are outlying from the main Andean chain (other examples from the Távara of species with this distributional pattern are *Hemitriccus rufigularis* and *Oxyruncus cristatus*). The record of Horned Curassow from the Távara suggests that the species may be, as was suspected, more continuously distributed between central Perú and central Bolivia. Nonetheless, the species clearly is rare, and is vulnerable both to hunting pressure and to forest disturbance.

Mammals of the Upper Tambopata/ Távara (L. Emmons and M. Romo)

Members of our expedition recorded 72 species of mammals in 19 days at our study sites on the Ríos Tambopata and Távara (Appendix 5). To these can be added eight other species from a list compiled by Walter Wust (1989). At least 23 more species are known from Explorer's Inn (L. Barkley, L. Emmons, and M. Romo, Appendix 6; T. Parker pers. comm.), bringing to 103 species the known mammal fauna of the Río Tambopata and its lowland tributaries. A number of other species, especially bats, a few rodents, marsupials, and carnivores, can be expected in the lowlands. The higher elevations of the Andean slope of the headwaters in Puno Department have an entirely distinct montane fauna that has been partially documented by earlier expeditions, whose results suggest that above 1000 m there may be some regionally endemic mammal species (Allen 1900, Osgood 1944, Cadle and Patton 1988). On this expedition we did not gain a high enough elevation to sample montane mammals.

All mammals recorded on the Ríos Tambopata and Távara are previously known from Madre de Dios, and there were no unexpected findings. None of the species is narrowly endemic to the region. However, there are high populations of species that have become rare from overhunting in many parts of Amazonia, especially tapirs and spider monkeys, but also jaguars, capybaras, white-lipped peccaries and the mid-sized monkeys. The river supports at least several groups of giant otters, as well as river otters. Woolly monkeys are uncommon or rare over most of their range and are threatened or extinct from hunting in many regions. In addition to numerous records of large mammals, our preliminary results show that the Tambopata basin has a high diversity of smaller mammals, such as rodents and bats.

The mammal species are unequally distributed in three major lowland habitat types: (1) river floodplain forests on recent alluvial soils; (2) *terra firme* forests on older terraces; and (3) bamboo forests. In numbers of mammal species that they include, the three habitats support decreasing numbers of species, with river floodplains being the richest habitats and large *terra firme* bamboo thickets the poorest. Each habitat is also a mosaic that varies in favorableness for mammals: floodplain that has standing water for much of the year is poorer than dryland habitat, while bamboo thickets intermixed with many trees or adjacent to forest include more mammal species than large pure stands, which are drastically depauperate.

Each habitat supports some species not, or rarely, seen in the others. The smaller primates and tree rats, for example, are most common in floodplain forest, while bamboo rats are largely restricted to bamboo, and armadillos are most numerous in *terra firme*.

The river supports at least several groups of giant otters, as well as river otters.

All habitats, including aquatic ones (rivers, oxbow lakes, swamps, streams) are needed to encompass and sustain the total species richness of the region. Indeed the mosaic habitat itself is a major source of the high species diversity.

There is a gradient in the influence of hunting along the river, with increasing densities of large mammals such as tapirs, capybaras, and spider monkeys, as well as large fish, with distance from permanent settlements. The upper drainage provides a renewal source for fish and game used by residents of the lower river and is thus important for maintaining the basic resources used by the local human community. The high price of gasoline is now the chief curb on subsistance and market hunting far upriver from communities, but this situation may not last.

Small Mammals

There was a high diversity of small mammals in the floodplain forest at the Ccolpa de Guacamayos. Bats were the most numerous group, with 25 species captured, and five more on the Río Távara (30 total). In the short time of our sample, we identified 12 species of small rodents (3 squirrels, 6 Muridae, and 3 Echimyidae), four opossums, and a rabbit. This should represent from 1/2 to 3/5 of the expected small mammal fauna.

Larger Mammals

The Río Madre de Dios is the southern limit of about four species of monkeys, so there are only nine species in the Río Tambopata basin, compared to 13 on the Río Manu. As noted above, the upper Tambopata and Távara basin has a dense population of spider monkeys. This favored game species seems in serious decline in many areas: it appears to be virtually extinct in much of the intensive Brazil nut and rubber exploitation regions of Pando (Bolivia), the lower Río Heath, and the lower

Río Tambopata (it is absent at Explorer's Inn) and from many of the more densely inhabited parts of Amazonia that we have visited. Woolly monkeys are even scarcer and have a patchy distribution in a smaller geographic range. The populations in the Tambopata-Candamo Reserved Zone therefore represent an important population reservoir for these species. The other large mammals that are scarce from overhunting in many parts of Perú, especially tapirs and capybaras, likewise have large populations on the upper Tambopata and Távara. The ease with which these mammals can be seen has made the Río Tambopata highly successful for ecotourism.

Herpetofauna of the Cerros del Távara (L. Rodríguez)

A partially different set of anuran species from that in the lowlands at the Ccolpa was observed ranging from 250 to 900 m (Appendix 8). At lower elevations, most species were the same as at the Ccolpa. Above 500 m, different species appeared: *Hyla callipleura*, an Andean hylid; an unidentified species of *Cochranella*; *Hemiphractus johnstonei*, a close relative of marsupial frogs, known from central Perú but not from Puno localities; and *Epipedobates* sp. nov., a poison dart frog similar to *E. petersi* and *E. macero* (Rodríguez and Myers 1993) that is now being described (Rodríguez and Myers, in prep.).

Also remarkable was the presence of some species previously reported from Cocha Cashu, a lowland site at 350 m, but not from other Madre de Dios localities; *Ischnocnema quisensis*, a species absent from Explorer's Inn and Cuzco Amazónico, and two new species of *Eleutherodactylus* (Rodríguez, in press; Rodríguez and Flores, MS) have been found on these ridges.

Most species inhabiting the ridges present specialized modes of reproduction.

... the upper Tambopata and Távara basin has a dense population of spider monkeys.

There are species with direct development (*Eleutherodactylus*), those carrying eggs throughout development (*Hemiphractus*), and also species with tadpoles having mouth parts highly adapted to running water (*Hyla callipleura*).

The hills here and in much of the lowland to the west are mostly covered with dense stands of bamboo whose origin and history is still a mystery.

Ccolpa de Guacamayos

Below the junction of Río Alto Tambopata and Río Távara, the river heads away from the mountains, carving its floodplain through a region of high terraces and lines of low hills. These hills of Tertiary age appear to be the eroded remnants of high alluvial terraces that have been recently uplifted. The final line of hills on the west side of the river is the site of the Ccolpa de Guacamayos tourist and research camp. Just upstream from the camp, the river is currently eroding these hills and exposing layers of salty clay. Great numbers of macaws, parrots, and other animals are attracted to this clay (or to some mineral or element that it contains).

The hills here and in much of the lowland to the west are mostly covered with dense stands of bamboo whose origin and history is still a mystery. The trail system of the camp traverses the hills as well as different aged terraces of recent river floodplain. Below camp to the north the river seems to be relatively stable, neither cutting nor depositing. This is probably because the ccolpa cliff deflects the force of the water. But past deposition has left what appear to be two different aged terraces, with very different forest composition. On the opposite side of the active braided river is a moderately high terrace (rarely flooded), with a large stretch of mostly continuous canopy old forest with many giant trees. The tree trunks usually have a thick cover of climbing plants suggesting that this habitat is considerably more moist than the lowland forests farther away from the mountains and hills.

We had only a couple of days to examine the plants of this area and our botanical report must be considered quite incomplete.

Forest Vegetation near the Ccolpa de Guacamayos (R. Foster)

Low Floodplain

Just above the rocky beach is a low terrace that is apparently frequently inundated and rather poorly drained. The 35-40 m canopy includes many large fig trees, including *Ficus perforata* and *Ficus killipii*, as well as large individuals with trunk diameters 50-100 cm or more of *Gallesia integrifolia*, *Calycophyllum spruceanum*, *Swartzia* (giant), *Enterolobium cyclocarpum*, *Sapium marmieri*, *Caryocar amygdaliforme*, *Clarisia racemosa*, and *Clarisia biflora*. This forest is estimated to be between 200 and 300 years old.

Common trunk climbers are of the genera *Rhodospatha*, *Philodendron*, *Monstera*, and *Asplundia*.

High Floodplain

The higher terrace adjacent to the low floodplain is apparently flooded only infrequently, i.e., once every several years. The 40-45 m high canopy is made up of huge trees characteristic of much of the older floodplain on fertile soils in the Madre de Dios drainage. The most common trees of this type are: *Hura crepitans*, *Ceiba pentandra*, *Ceiba samauma*, *Poulsenia armata*, *Dipteryx micrantha*, *Luehea cymulosa*, *Ficus trigonata*, *Spondias mombin*, *Apuleia leiocarpa*, *Sloanea guianensis*, and *Otoba* sp. There are also a few *Hymenaea courbaril*, a species more characteristic of ancient, unflooded terraces.

The characteristic medium-sized trees of the understory are *Naucleopsis ulei*, *Lunania parviflora*, and *Iryanthera juruana*. Characteristic shrubs and treelets are *Carpotroche longifolia, Chrysochlamys ulei, Stylogyne cauliflora, Piper laevigatum, Psychotria viridiflora, Clidemia septuplinervia, Calyptranthes* (2 spp.), *Rudgea,* and *Alibertia pilosa.*

Palm Swamp (Aguajales)

To the west of the camp at the base of the hills is a swamp forest that is seasonally flooded and always wet, characterized by the palm *Mauritia flexuosa* (aguaje). This aguajal, like those of Explorer's Inn (but unlike those of Manu), also has an abundance of the tree *Luehopsis hoehnei.* This is only the second report of this species from Perú. From the satellite image it is clear that to the west of this site along the base of the hills, there are more and larger aguajales, presumably of similar composition. The origin of these swamps is not immediately obvious, but may have a simple explanation in the relation of the river to the hills.

Hills

Although the hills to the south of camp are mostly dominated by dense stands of the bamboo *Guadua weberbaueri,* there are scattered trees (30-35 m) throughout, and small continuous patches of forest in ravines and on some slopes of the hills. Some characteristic high floodplain trees such as *Apuleia leiocarpa* and *Dipteryx micrantha* are also common here. But there is a suite of tree species that characterize this habitat throughout the region, such as *Cedrelinga, Huberodendron, Couratari,* and *Dussia,* with *Eschweilera, Tapirira, Byrsonima,* and *Sparattosperma* in the drier areas. We did not encounter any Brazil nut (castaña) trees. Common subcanopy species and genera include *Senefeldera, Cheiloclinium, Iryanthera, Wettinia, Inga* spp., *Diospyros, Naucleopsis* [short], *Quararibea ochrocalyx, Tabernae-montana, Coccoloba mollis, Miconia,* and Myrtaceae spp. Common shrubs are *Manihot* sp., *Clibadium,*

Oxandra, Chelyocarpus, and dense stands of *Geonoma deversa.* A distinctive and common herb is *Monotagma.* Common lianas are (Bignoniaceae) spp., *Mendoncia* spp., *Dioscorea,* and *Salacia.*

Bamboo

In the lowlands were two species of large bamboo, *Guadua angustifolia* and *G. weberbaueri,* sometimes occuring in mixed clones. On the hills we found only *G. weberbaueri.*

A stem of *Guadua weberbaueri* grows until it can no longer support its weight and then falls, to be replaced by sprouts on the lower half of the stem and from the ground by other ramets of the clone. This happens sooner for the small clumps in small forest gaps. When growing in great stands, the stems are held up by other shoots in the clone or adjacent clones, and the stems reach greater age and diameter. This causes a great build up of weight 10-20 m above ground, which then falls en mass bringing down both young and old stems in an impenetrable jumble. Where there is a thin tree canopy and the bamboo sparse below, there is a much more dense understory and ground layer of other plant species. When the canopy is open and the bamboo thick, there is very little understory.

At the time of our visit, *Guadua angustifolia* was in the process of flowering and dying. This process appears to take place over a few years. A flowering shoot apparently does not die immediately near the base. This allows sprouts from the lowest nodes to grow up to considerable size before flowering themselves within a year or so, keeping the process going with sprouts of the sprouts. But each succeeding effort is smaller than the one previous until all main stems and sprouts have flowered and the clone finally dies. The dying and dead clones are replaced by seedlings of their own or by *Guadua weberbaueri* clones that have been suppressed underneath.

Birds of the Ccolpa de Guacamayos
(T. Parker, A. Kratter, T. Schulenberg and W. Wust)

As one would expect, the avifauna of the Ccolpa is very similar to that of the Explorer's Inn Reserve farther downstream, although fewer species have been recorded at the Ccolpa (Appendix 2). Many of the "missing species" are birds associated with forest streams, lakes or their margins, all of which are habitats that are absent or peripheral near the Ccolpa; are species that are migrants from the north temperate regions, and would not be expected to occur at the Ccolpa during the periods it was visited; or, are relatively uncommon species that presumably are present, but to date have escaped detection.

The Ccolpa's closer proximity to the Andean foothills is reflected in the presence of, or great abundance of, primarily foothill species, such as *Synallaxis cabanisi, Myrmotherula longicauda, Mionectes olivaceus* and *Elaenia gigas*. It is also interesting to note the presence of populations of some species (particularly *Crypturellus atrocapillus, Pyrrhura picta*, and *Picumnus rufiventris*) that are absent, or very scarce, at the Explorer's Inn Reserve, but which are of regular occurrence at Cocha Cashu on the Río Manu.

Large numbers of parrots and macaws visit the clay lick at the Ccolpa. Also notable are the very extensive stands of bamboo in this area. Populations of birds that are closely associated with bamboo (Parker 1982) are larger here than at the Explorer's Inn Reserve; many of these are species with restricted distributions, e.g., *Simoxenops ucayalae, Automolus dorsalis, Cymbilaimus sanctaemariae, Drymophila devillei, Cercomacra manu, Poecilotriccus albifacies* and *Ramphotrigon fuscicauda*. The area around the Ccolpa also supports what may be the largest population density of the extremely scarce antbird

Formicarius rufifrons (Kratter in press), whose known distribution lies entirely within the Department of Madre de Dios.

Herpetofauna of the Ccolpa de Guacamayos (L. Rodríguez)

In eight days of sampling at the Ccolpa, we recorded a total of 44 anuran species, 12 lizards, and ten snakes (Appendix 8). The most striking difference between the anuran faunas of the Ccolpa and Explorer's Inn, only 40 km apart, is in the number of species of *Eleutherodactylus*. While only seven species have been recorded at Explorer's Inn after at least four seasons of fieldwork, nine species have already been found at the Ccolpa in only eight days of collecting. It appears that the greater number of *Eleutherodactylus* at the Ccolpa is due to its proximity to the first Andean ridges. The base of the Andes is characterized by the highest co-ocurrence of species in the genus; many of the species do not penetrate into the Amazonian lowlands. From Table 3, it is obvious that the sampled areas and Cocha Cashu share most species of *Eleutherodactylus* and can be placed in the same biogeographic assemblage.

Colostethus marchesianus, a small, ground-dwelling dendrobatid that is widespread and abundant in other lowland localities, is one of the more notable species absent at the Ccolpa. However, it was rather abundant on the ridges of the Távara above 450 m. I suspect that the discrete distribution of this frog might be related to some specific microhabitat or microclimatic factor more than to species interactions, because it is known to coexist at the Explorer's Inn Reserve and other localities with the same set of terrestrial and diurnal species present at the Ccolpa.

Also remarkable is the unusually low number of small treefrogs (*Hyla*). While small

Large numbers of parrots and macaws visit the clay lick at the Ccolpa.

Table 3. Species of *Eleutherodactylus* in different localities of Madre de Dios and Puno and number of species shared between sites.

	Távara	Ccolpa	Tambta	CA	CC
Távara	8	5	2	4	6
Ccolpa		9	5	6	8
Tambopata			7	4	4
Cuzco Amazónico				7	5
Cocha Cashu					12

Total number of species = 14

hylids are represented by around seven species at other sites in Madre de Dios, only one of those, *Hyla parviceps*, was found at the Ccolpa. The reason for this difference may be related to the season (end of the rainy season). In a collection of frogs made at Cocha Cashu in May 1988, the number of small hylids was also very low (Rodríguez, unpub. data). At the Ccolpa, although rains were still constant, most forest ponds were drying out and very few males were heard calling. Abundant juveniles of *Physalaemus petersi* and *Leptodactylus leptodactyloides*, of a certain size, also indicated the end of the reproductive season.

Frogs at the Ccolpa showed habitat specificity. Terrestrial frogs were highly abundant in the floodplain, with several species of *Eleutherodactylus* occupying the low vegetation. In the bamboo and upland terraces, frogs were generally less abundant, but *Epipedobates trivittatus* and *Dendrobates biolat* were especially common. The occurrence of *Leptodactylus knudseni* in the Ccolpa floodplain and of *L. pentadactylus* across the river might be interpreted as habitat segregation between these two species, also observed at Cocha Cashu.

Some species were apparently reproducing during the sampling period. A big treefrog, *Phyllomedusa bicolor*, was often calling along the pond in the floodplain and tadpoles of *D. biolat* were found inside bamboo. On the river bank, *Hyla boans*, *H. geographica*, *Bufo poeppigi*, *Physalaemus petersi*, species that are known to reproduce during the dry season, were calling around pools indicating the beginning of the reproductive season.

Among the more relevant new reports for Madre de Dios are those of *Osteocephalus* cf. *pearsoni*, also present at the Pampas del Heath (Icochea, this report) and *Chiasmocleis bassleri*, for which it probably is the southernmost locality. A probable new species of *Eleutherodactylus* (*E.* cf. *cruralis*) was collected at the Ccolpa; it has a very characteristic call and some morphological features that do not place the species in the *E. fitzingeri* or the *E. unistrigatus* groups, to which all other species in the area belong.

Explorer's Inn Reserve

Only one other site in the Peruvian Amazon, the neighboring lower Manu River, can rival the Explorer's Inn Reserve (formerly called the Tambopata Reserved Zone or Antigua Zona Reservada de Tambopata) for the amount of information about its biota. With its diverse network of trails, comfortable accomodations, farsighted resident naturalist program and subsidies for Peruvian biologists, this tourist lodge has become a major center for tropical research.

A principal factor in the attraction of this 5500 ha area is the diversity of lowland topography that it encompasses. It includes ancient, high, sandy terraces, as well as low recent clayey floodplain and many terrace steps in between. It has a large oxbow lake of the Río Tambopata as well as small oxbow lagoons of the Río La Torre tributary. Each of these habitats has its own set of common and rare species, giving the area as a whole a very high biological diversity.

Our current rapid assessment did not include a survey of Explorer's Inn Reserve (except for additional butterfly work) for the obvious reason that it has already had fairly intensive long-term assessment. We have decided instead to include brief comparative summaries of its biota here, and include summary lists of species in the appendices of this volume (Appendices 3, 6, and 12) and in the botanical supplement. Some of these lists have or will be published elsewhere with more annotation. The bird list, however, is the first complete list to be published for this site and represents many years of research, mainly by Ted Parker. The mammal list is also published here for the first time, but as it is based on only brief inventories, it is incomplete.

Summary of the Vegetation of Explorer's Inn Reserve (O. Phillips)

Most of the area of the Explorer's Inn Reserve appears to be on Holocene or Quaternary floodplains of the Río Tambopata, and only a small fraction is within the contemporary floodplain. Nine forest-types are described in a recent extensive study (Phillips 1993b).

Because of the complex flooding regimes along the Tambopata and La Torre rivers, three different categories of contemporary floodplain forest are recognizable: lower, middle, and upper floodplain forest. The categories reflect the relative probability of different locations in the floodplain being flooded, and the length of time that the vegetational succession has had to develop since the location was part of the river channel.

Increased time since the most recent river flooding correlates with increased acidity, and therefore lowered availability of nutrients. However, even the floodplain soils are not especially fertile—apparently because of the influence of the Río La Torre, which mostly drains weathered ancient Tertiary sediments and therefore contains a nutrient-poor sedimentary load. The relative poverty of the floodplain included in the reserve may make it less typical of the floodplains of the major rivers in the area, but perhaps more representative of much of south-central Amazonia.

Lower Floodplain Forest: Young successional forest, usually less than 20 m high, developing from open river beaches. Dominated by *Cecropia membranacea*, *Sapium* spp., *Croton lechleri* (though apparently not in the La Torre floodplain), and juveniles and young adults of *Ficus insipida*. The only conspicuous palm is *Socratea exorrhiza*. Small lianas such as *Paullinia alata* form locally dense tangles. Large (3-4 m), often clonally spreading herbaceous plants are dense in the understory, e.g., *Heliconia* spp., *Calathea* spp., *Renealmia* sp., and *Costus* spp.

Middle Floodplain Forest: Flooded between once a year to once a decade. Some trees reaching 35-40 m, especially *Calycophyllum spruceanum*, *Ficus insipida*, and *Cedrela odorata* (rare now). Palm trees dominate middle layers and regenerate profusely, especially *Iriartea deltoidea*, *Socratea exorrhiza*, *Astrocaryum gratum*, and *Attalea* sp. It has the same dense herbaceous understory as the preceding forest type.

Upper Floodplain Forest: Flooded very

briefly and rarely, probably less than once a decade, and mostly well-drained. The canopy is about 30 m high. Most *Calycophyllum spruceanum* and *Ficus insipida* are dead by this stage. Prominent canopy trees are *Ceiba pentandra*, *Chorisia* sp., *Ficus* spp. (stranglers), *Pouteria* spp., *Pourouma* spp., *Brosimum alicastrum*, *B. lactescens*, and *Jacaratia digitata*. The common palms are the same as for the preceeding type. Dense clones of large monocotyledonous herbs are infrequent.

Lower Previous Floodplain Forest (= "Old" Floodplain Forest): Former floodplain that is no longer inundated, but canopy consists of trees that probably became established when the area was still occasionally flooded, i.e., within the last 200 years. Mostly high, closed-canopy forest with an open understory. Floristically, this forest type is intermediate between Upper Floodplain Forest and Upper Previous Floodplain Forest. Conspicuous trees include giant individuals of *Spondias mombin*, *Hymenaea* sp., *Dipteryx micrantha*, *Ceiba pentandra*, and *Pouteria* sp. Two palm tree species so abundant in the previous two forest types are only occasional in this forest (*Attalea* sp. and *Astrocaryum gratum*).

Terra Firme Clay Forest (Upper Previous Floodplain Forest): This habitat, on which the station buildings are located, is up to 20 m above present floodplain forests. Although it is the most frequent forest type in the Explorer's Inn Reserve, it may be very rare in the Madre de Dios area. The fluvial origin is confirmed by the occasional swampy remnants of oxbow lakes, but it probably has not been flooded by the Río Tambopata for more than 1000 years. The canopy is more broken up than in the above forest types and the understory and ground layers are correspondingly dense. Prominent trees include *Parkia* spp., *Tachigali* spp., *Inga* spp., *Pourouma minor*, *P. guianensis*, *Pseudolmedia macrophylla*,

P. laevis, and *Bertholettia excelsa* (castaña). Palms are prominent, especially *Iriartea deltoidea*, *Euterpe precatoria*, and *Oenocarpus mapora*. The shrub *Psychotria poeppigiana*, with its conspicuous red "lips," is prominent along the trails.

Terra Firme Sandy-Clay Forest: Ancient alluvial terraces or low hills of eroded terraces. The soils are sandy, well-drained, highly leached, acidic, and usually with a substantial layer of surface raw humus. *Huberodendron swietenioides* and *Cedrelinga cateniformis* are emergent trees, and *Bertholettia excelsa* is relatively frequent. Other important trees are *Hevea guianensis*, *Tachigali* spp., *Pourouma minor*, as are small trees such as *Bixa arborea*, *Diospyros melinonii*, *Ouratea* sp., and an unusual diversity of Melastomataceae. The palms *Oenocarpus bataua*, *O. mapora*, and *Oenocarpus* sp. nov. (?) are all quite frequent. A variety of the other usual palms are present, but at much lower densities than in the other forest types.

Terra Firme Sand Forest: Encountered off the reserve on the Río Tres Aguas Negras tributary of the Río La Torre. An extreme version of the last forest type with much more sand in the soil. The forest is reduced in stature (15-20 m high), with a high density of small stems. *Hevea guianensis* is common, while palms are infrequent. This forest type needs much more study.

Permanently Water-logged Swamp Forest: Flooded or water-logged sections of filling-in oxbow lakes, characterized by either *Mauritia flexuosa* palms, by trees of *Lueheopsis hoehnei*, or by both together. Very few species in the region can tolerate the prolonged immersion in the still, anaerobic water of these swamps.

Seasonally Water-logged Swamp Forest: Poorly drained forest with characteristic "channel/hump" (or "ditch/hummock") formations. The humps are colonized by plants, with channels between them filled with water intermittently. Possibly derived from *Mauritia* palm swamps after many centuries. Mostly covered sparsely with small trees, mainly palms, especially *Oenocarpus bataua*, *Euterpe precatoria*, and *Attalea phalerata*. Other trees include *Symphonia globulifera*, *Licaria armeniaca*, *Crudia glaberrima*, and *Maquira coriacea*. The understory and ground layer is very dense and has numerous vegetatively-spreading palms such as *Bactris* spp. and *Oenocarpus mapora*.

Summary of the Floristics of Explorer's Inn Reserve (C. Reynel and A. Gentry)

There are approximately 1400 species of vascular plants known from the Explorer's Inn Reserve. This number will increase as more collections are made, especially from further quantitative samples of the vegetation. For the flora as a whole, the two largest families of plants are the Leguminosae (106 species) and the Rubiaceae (101 species). This is typical for much of Amazonia. Continuing in order are the: Bignoniaceae (58 spp.), which is also the largest family of lianas, Moraceae (55 spp.), and ferns and their allies (54 spp.). These are also always among the most speciose families of Amazonia, but the Bignoniaceae rank higher at this site than anywhere else for which comparable data are available. The other large families are Euphorbiaceae, Melastomataceae, and Piperaceae (44 spp. each); and Araceae, Palmae, Solanaceae, Gramineae, Orchidaceae, Annonaceae, Lauraceae, and Sapindaceae (28-33 species each). This familial composition is typical of the Upper Amazon.

The largest genera in the flora are *Piper* with 34 species and *Psychotria* and *Inga* with 32. An estimated 12% of the species in the flora are endemic to the southwestern part of Upper Amazonia (i.e., Acre in Brazil, Madre de Dios in Perú, and parts of La Paz and Pando in Bolivia). It thus represents a distinct subset of the Amazon flora.

Each of the mature, noninundated forest types have from 155-180 species of trees (≥10 cm dbh) per hectare (Gentry 1988a, b). This is probably close to average for Amazon Basin mature forest, but is substantially higher than in neighboring regions at the same latitude (Amazonian Bolivia and southern Amazonian Brazil). The families with most species of trees are in descending order: Leguminosae, Moraceae, Lauraceae, Annonaceae, and Sapotaceae. Eight of the ten most speciose families are shared with plots at La Selva Station, Costa Rica, but only four are shared with plots in Ducke Reserve in central Amazonian Brazil. The genera with most species per plot are *Inga*, *Pouteria*, *Pourouma*, *Virola*, and *Neea*. The species with most individuals per plot are *Iriartea deltoidea*, *Pourouma minor*, *Iryanthera juruensis*, *Leonia glycycarpa*, *Iryanthera laevis*, *Pseudolmedia laevis*, *Raucheria punctata*, *Rinoria viridifolia*, *Oenocarpus bataua*, and *Socratea exorrhiza*.

Except for the Melastomataceae and Lauraceae, there is a remarkable similarity in the relative species representation of each family from plot to plot. Some species are heavily concentrated in a single plot, such as *Lueheopsis hoehnei*, *Rinorea viridifolia*, *Ampelocera verrucosa*, *Euterpe precatoria*, *Leonia glycycarpa*, *Pseudolmedia laevigata*, *Roucheria punctata*, and *Virola sebifera*, but most species have significant representation in more than one plot. Many of the differences in species abundance between plots appear to be caused by soil differences or some other physical difference in the habitat such as flooding or shading. For many species, the cause may be more related to chance historical factors. This will be better understood if more plots are replicated in the same habitat, but in different locations in the Tambopata-Candamo Reserved Zone.

There are approximately 1400 species of vascular plants known from the Explorer's Inn Reserve.

Importance of the Explorer's Inn Reserve Permanent Plots (A. Gentry)

Because of their overwhelming diversity, tropical forest plants are very difficult to identify. One cannot just go out in the forest and identify plants as one comes to them. Thus for all but a few experts, the tropical forest constitutes a kind of taxonomic black box intractable to studies that demand knowledge of species identifications. Moreover, even taxonomic specialists intent on cataloging the plants of a tropical rain forest site usually limit themselves to plants they can find in flower or fruit, which means that hard to see and hard to collect canopy trees and lianas are often neglected.

The most effective way to overcome these problems is to set up large permanent plots in tropical rain forest with numbered trees that can be relocated after they have been identified. One hectare plots for censusing trees and lianas ≥10 cm dbh are the most common sample size and have been called the "industrial standard" of tropical forest inventory. These are the plots that have been established at Tambopata. A huge amount of effort goes into establishing such plots, since each tree must be identified to species, usually involving climbing it to obtain a sample of leaves, then trying to match the voucher with identified fertile specimens in one of a handful of large herbaria with major tropical collections. Until recently, the taxonomic expertise necessary to field-identify the plants to family and genus, a necessary pre-condition to herbarium specific identification, did not exist.

Inventorying such plots has forced taxonomists to deal with a multitude of non-fertile plants they would otherwise ignore. This has led to the discovery that there are still huge numbers of undescribed tree species in Amazonian forests. For example, in complex large families like Lauraceae and Sapotaceae

there are dozens of apparently undescribed species in the Tambopata plots. Examples of new species from those plots include *Sarcaulus*, *Tabebuia incana*, *Mezilaurus*, and *Aspidosperma tambopatensis*. Even some of the most common species turn out to be undescribed, like *Astrocaryum gratum*, one of the most abundant trees in the alluvial plot. Dramatic disjunctions and range extensions are also common. An example from Tambopata is *Dialypetalanthus fuscescens* (Dialypetalanthaceae), with a single tree in plot 3, which represents a new species, genus, and family record, not only for Amazonian Perú, but also for upper Amazonia. *Lueheopsis hoehnei*, the most common species in the swamp plot, is a new record for Perú. Thus, these plots are very important for the basic documentation of the flora. Since the trees are numbered and can be relocated, the Tambopata plots constitute a kind of living gene bank for many rare species that can be easily studied, or their seed collected, no where else in the world.

One unique aspect of the Tambopata plots is that they sample a wide variety of habitat types—swamp forest, alluvial forest, lateritic soil forest, and sandy soil forest. This is the only place in the world where such a series of plots on different substrates exists. Thus the Tambopata plots provide a unique record of how plant species change between habitats, with climatic and geographic factors controlled for. This was a key element in developing the habitat mosaic theory to explain Amazonian diversity (Gentry 1988b, 1989).

Vegetation dynamics are very important to ecological theory and to potential management of tropical forest. For example, tree fall gaps turn out to have major importance in understanding tropical forest community dynamics. New ideas for sustainable use of tropical forest depend directly on understanding how dynamic the natural forest

... there are still huge numbers of undescribed tree species in Amazonian forests.

New ideas for sustainable use of tropical forest depend directly on understanding how dynamic the natural forest is.

is. Such phenomena occur on time scales of decades or centuries and can only be understood with long term plot data, and few Neotropical tree plots are old enough to provide data. Incredibly, the only Amazonian plot for which ten years of data on mortality and recruitment has been published is at the Cocha Cashu field station in Manu National Park. Plot 1 at Tambopata has now been in existence for fifteen years and is the second oldest floristically inventoried plot in Amazonia for which data on stand dynamics are available, and the other Tambopata plots, with 11 years of data are the runners-up. An example of the importance of such data is the new discovery that the single best predictor variable of tropical forest community diversity is the turnover rate of the forest (Phillips et al. 1994a). This discovery, partly based on the Tambopata plots, provides for the first time a credible explanation for the species-energy relationship in the tropics. Higher turnover occurs in more productive ecosystems where trees grow and die most rapidly. This in turn reduces the chances of competitive interactions between ecologically equivalent species, and results in higher diversity as different ecologically equivalent species successfully colonize different gaps. One of the Tambopata plots is the second most dynamic ever measured (the most dynamic is near Iquitos), and all but the swamp plot are highly dynamic. Apparently the relatively rich soils, high rainfall, and local habitat mosaic of upper Amazonia combine to make these forests among the most dynamic and species rich in the world.

Perhaps the greatest importance of the Tambopata plots is as a unique resource for other investigators. For example, the BIOLAT sampling scheme, first developed with the Tambopata plots, focuses on collecting and observing insects, birds, and other animals against a background of identified plants so that their distributions and behaviors can be

more fully understood. A good example of the importance of the Tambopata plots is in ethnobotany where the doctoral thesis of Oliver Phillips (1993b) was only possible due to the existence of these plots. Phillips quantified the usefulness of the forest and of different species and different forest types to local people at Tambopata, in the process essentially establishing ethnobotany as a true science based on hypothesis testing (Phillips and Gentry 1993a, b; Phillips et al. 1994b). Going with his informants around the tree plots, he was free to concentrate on ethnobotany because the plant identification had already been completed. This study gives these plots historical significance. It also points out new ideas of economic and political relevance. For example, the richer-soil mature floodplain forest, conservationally neglected and by far the most threatened habitat in the Puerto Maldonado area, is the most useful forest to the local people and would be a key element in any attempt to establish extractive reserves.

Another new idea, with potential global political significance that has come from analysis of the Tambopata tree plot long-term turnover data and data from a select group of tree plots from other tropical forests in the world, is that tropical forests seem to be growing faster in recent years. This may partly explain what has happened to the "missing" CO_2, as the atmospheric buildup of this critical greenhouse gas lags behind the rate of buildup anticipated from the burning of fossil fuels and tropical forests. Speeded up growth of tropical forests, fertilized by extra CO_2, may be taking up a substantial quantity of the atmospheric CO_2. This discovery is important because of the widespread ramification of greenhouse warming and also because it suggests that the dampened greenhouse heating experienced so far may be temporary, and that the rate of atmospheric CO_2 increase and greenhouse warming may increase as the currently faster-growing forests die faster as well, so releasing more carbon into the air. Moreover, cutting of additional forests removes this planetary safety valve. The great length of the Tambopata forest census data relative to other sites in Amazonia, makes it critically important to maintain these plots to test the future direction of this and other changes. These issues are explored in two recent papers in *Science* that attracted international media coverage (Phillips and Gentry 1994, Pimm and Sudgen 1994).

Species of Economic Interest in the Flora of Tambopata and its Importance as a National Heritage (C. Reynel and O. Phillips)

Few countries have been as favored in terms of distribution of biological diversity as Perú. The great quantity of species inhabiting Perú has resulted in the existence of many native varieties of plants of great economic significance.

Currently, a common subject for discussion is the loss or erosion of genetic resources resulting from the massive destruction of species diversity. A good example of this situation has been documented in the case of different wild varieties and species related to the potato (*Solanum tuberosum*), which is a native plant of Perú and one of many crops that originated in the Andean region (Ochoa 1975).

Until three decades ago, a large number of genetic varieties of potatoes grew naturally in Perú. Thus Perú has been the world's natural storehouse for potatoes. Now many of these plant populations have been eradicated and extinguished—for reasons that are increasingly frequent, the destruction of natural vegetation by unplanned urban and rural expansion.

One dramatic consequence of this situation concerning crops or their relatives is that the loss of genetic variation results in the

lack of tools to combat the natural enemies of the cultivated crop. For example, each genetic strain possesses a different level of resistance or adaptability to certain conditions such as insect pests or adverse climate changes.

Another consequence of this problem is that the loss of varieties and species is resulting in Perú's transformation from exporter of natural resources to importer. The export of these potato crops are of extreme importance to Perú's economy and development.

The example of the species and varieties of *Solanum* in Perú is illustrative of how Peru's natural heritage and genetic resources should be protected. Protected areas should be created to preserve Peru's valuable species and varieties.

The Tambopata-Candamo Reserved Zone (TCRZ) contains a great quantity of plant species that are of economic interest. Many of them are endemic to southern Perú and their natural populations are at risk of destruction if deforestation rates intensify. Thanks to both Peruvian and foreign research efforts, we now have an approximate notion of what flora exists in this part of Perú and understand the importance and role of these species in the economy of local communities (Gentry 1989; Phillips 1992, 1993a; Reynel and Gentry, in prep.).

The TCRZ harbours high concentrations of wild tropical species and genetic strains or varieties that are of great economic importance. For example, recent studies reveal that in the case of fruiting trees, the lowlands of the reserve contain eight wild species of both regional and international importance, and at least 14 species that are congeneric wild relatives of tropical fruits important to the international economy. We mention the genera and common name of these cultivated species: *Annona* (chirimoya, guanábana), *Bactris* (pejibaye, pijuayo), *Bertholletia* (castaña),

... loss of varieties and species is resulting in Perú's transformation from exporter of natural resources to importer.

Carica (papaya), *Eugenia* (pomarrosa), *Inga* (pacae), *Juglans* (nogal), *Mauritia* (aguaje), *Morus* (morera), *Paullinia* (guaraná), *Persea* (palto, aguacate), *Pourouma* (uvilla), *Pouteria* (lúcuma, caimito), and *Rollinia* (anona).

The area also contains a large number of plants with medicinal properties of economic interest, only a fraction of which have been studied. Additionally the native communities who live in the region are a reservoir of knowledge concerning these species, which constitute one of the most important potential resources of the Amazon region. Three examples of this are: (1) a liana *Uncaria* (uña de gato), whose bark has demonstrated properties promising in the treatment of some types of cancer; (2) the trees *Croton draconoides* and *C. lechleri* (sangre de grado), whose resins contain healing agents employed for the treatment of ulcers; and (3) the well known *Ficus insipida* (ojé), whose latex produces a medicine used in the treatment of intestinal parasites. The list of plants with medical uses is long (Phillips and Gentry 1993a, b).

Although our knowledge of the flora in the area is still preliminary, we know that at least 223 of the tree species in the zone are used in woodworking; among those are a few dozen species with very high quality wood.

The following is a list of some of the most well known species of economic interest with brief comments. All of the species included below are well represented in the region, and the protection of this zone would be an effective means of conserving them. It is necessary to indicate, however, that the knowledge of the flora in this region is still incomplete; in particular, the upper watersheds in the area have only been studied in a preliminary fashion, and without doubt they harbor a great number of species of scientific and economic interest.

The area ... contains a large number of plants with medicinal properties of economic interest, only a fraction of which have been studied.

Bertholletia excelsa (castaña; Brazil nut)

A tree of great size, reaches 35 or more meters in height. It grows in low to medium altitudes in the reserved zone. The edible part is the seed that is found in the interior of the round fruit, which is 12-18 cm in diameter. This seed has an exceptional protein content. It is extracted artfully from the fruit and can be consumed immediately. It is used often in confectionary and has a good international market. The castaña has generated an important economic movement in the selva sur of Perú, and is the base of the economy for hundreds of families in the region.

Theobroma cacao (cacao)

A small tree with fruit that grows on the trunk and branches. The seeds of this plant, which grows wild in the area, is dried and later made into chocolate, a product with a wide regional and international market. It grows in low to mid elevations in the area, and is confined to the endangered rich soil forests of the upper floodplain and old floodplain.

Hevea guianensis (caucho, jebe, shiringa)

A very large tree with tricapsular fruits. It is found in low to mid elevations in the zone. From the incisions on the trunk, rubber is extracted, which flows naturally and is collected in buckets and is later refined. The international market for rubber has varied throughout recent history due to the production of synthetic rubber, but natural rubber continues to be required for specific uses. Trees are not currently tapped in the region, but many larger *Hevea* trees bear the scars of rubbertapping that occurred as recently as 1970.

Cedrela fissilis (cedro)

A very large tree with big capsular fruits. It grows throughout the area of the reserve. *Cedrela fissilis* is closely related to *C. odorata*, a species highly prized for its valuable wood, and used for some of the finest carpentry work. It is of great demand in both regional and international markets.

Bixa arborea (achiote de monte)

A tree of small to medium size, with capsular fruit and reddish seeds. This is a close relative to cultivated achiote, or anato (*Bixa orellana*), extensively cultivated in some areas of the Amazon, whose seed is used as a natural dye with a growing international market.

Mauritia flexuosa (aguaje)

A palm which reaches 25 m in height, with oblong fruits with oily, edible yellow pulp. This species is valued in the Amazon region for its fruit, which is eaten directly from the tree, or in the form of soft drinks and frozen ice creams. The fruit has good potential for industrialization and exportation. However, the population of this species is rapidly deteriorating since it is often cut in destructive harvesting of the fruit. In the reserve, it is common in inundated areas.

Ficus insipida (ojé)

A non-strangling fig tree, often reaching 30 m high with a trunk that yields abundant white latex when cut. The latex is extracted in small amounts, and is used fresh or processed industrially for medical treatment of intestinal parasites; it has a regular demand in the regional market and has been exported as well. It grows in abundance in the lowlands and floodplains of the reserve, but the population of mature trees is threatened by expanding agricultural development of riverbank forests.

Annona hypoglauca (anonilla)

A small to medium-sized tree with fruits that are edible and similar to guanábana (*Annona muricata*) and the "anona," or cultivated "chirimoya" (*A. cherimolia*), to which it is closely related. The latter has a growing regional and international market. *Annona hypoglauca* is common in the low to mid altitude of the reserve.

Cedrelinga cateniformis (pino peruano, tornillo)

A very large tree with elongated and twisted fruit; its wood is very workable, it is durable and very much appreciated for use in fine carpentry. This species has a national market. It is naturally common in poorer soil forests of the low to mid elevations of the reserve, but small-scale selective logging has targeted this species. Without conservation measures this pressure could endanger the species, much as other timber species such as *Swietenia macrophylla* (mahogany) and *Cedrela odorata* (cedro) have become endangered.

Podocarpus sp. (romerillo, inimpa)

A very large tree with a very straight shaft; its wood is of excellent quality, durability, and has fine grain very appreciated for fine carpentry. It has represented one of the more interesting alternatives to reforestation in the zones of la Selva Alta in Perú. The "romerillo" is present in the high elevations of the reserve. These species are highly threatened by extinction in Perú.

Birds of the Explorer's Inn Reserve (T. Parker and T. Schulenberg)

The lowlands (below ca. 500 m), including the meandering floodplains and older terraces farther back from the rivers, of southwestern Amazonia support some of the richest plant and animal communities on earth. Additionally, almost all of the most geographically restricted bird species in this region are found only in floodplain forests along the rivers— a pattern that will no doubt be found in other groups as well. These forests are also

... almost all of the most geographically restricted bird species in this region are found only in floodplain forests along the rivers

extremely vulnerable to human activities such as logging, gold mining, hunting, and agriculture.

Almost 575 bird species have been recorded in the Explorer's Inn Reserve, all within an area of approximately 5500 ha, since ornithological investigations began at this site in the late 1970's (Appendix 3). This is comparable to the totals recorded at the Cocha Cashu Biological Station in Manu National Park, not far to the northwest (Terborgh et al. 1984, Terborgh et al. 1990). This extraordinarily high total results from several favorable factors, not least among these is the high visitation rate by ornithologists and birdwatchers. Although previously unrecorded species are detected annually, the composition of the resident avifauna is as well known as at any other site in Amazonia. It is quite rare now that a resident, or potentially resident, species is added to the site list; most additions to the list are of birds that are obviously migrating or dispersing individuals. Indeed, approximately 10% of the species recorded at the Reserve are no more than vagrants, and are not regular members of the bird species community.

More important biologically is the habitat diversity of the Reserve. Even within the relatively small size of the area surveyed, several distinct types of forest can be found. As mentioned above, this diversity is due to the meandering actions of the Río Tambopata and its tributaries; the importance of such rivers in creating and maintaining the landscape has been well-described elsewhere (Terborgh et al. 1984). Forests in the river floodplain, which are regularly subjected to flooding by the river, are particularly rich in species. Almost 50% of the forest bird species are more or less restricted to this lowland forest; less than 10% of the forest bird species are similarly restricted to *terra firme* forests on higher ground that never experience flood-

ing. Thickets of bamboo in several areas of floodplain forests on the reserve form a microhabitat very important for several species of birds, many of them rare (Parker 1982). The few islands in the relatively small Río Tambopata and its tributaries do not have a distinctive bird community, as is the case with larger rivers farther north in Perú (Rosenberg 1990). Nonetheless the successional vegetation along even these rivers provides habitat for a distinctive bird community (Remsen and Parker 1983). Oxbow lakes and associated marshes also provide essential habitats for resident and migratory waterbirds.

Approximately 80 species recorded here are known or strongly suspected to be migrants to Amazonia. Of those species whose geographic origin can be inferred, equal numbers are of austral (37) as opposed to north temperate (35) origin. It is worth noting, however, that at least eight of the north temperate migrants are shorebirds that occur in the reserve strictly in passage, rather than overwintering in Amazonia. Most species of migratory bird at the reserve, regardless of geographic origin, primarily utilize more or less open habitats, such as river or forest-edge, or lake margins, rather than the forest interior. Regular migrants that extensively utilize forest include *Coccyzus americanus* (north temperate), *Myiopagus viridicata* (austral), *Myiarchus swainsoni* and *M. tyrannulus* (both austral), *Myiozetetes luteiventris* (from Middle America), and *Vireo olivaceus* (austral); the majority of even these migrant species are not restricted to forest, however, but also occur in "zabolo" or other successional habitats.

The Explorer's Inn Reserve is not a pristine forest, as at Cocha Cashu. Macaws are still relatively common at the reserve. Populations of some other large species, such as Razor-billed Curassow (*Mitu tuberosa*), are very low, however, reflecting past hunting pressure.

Notable at the Reserve are the diversity of parrots (19 spp.), antbirds (52 spp.) and tyrant flycatchers (90 spp.). Although cotingas are not particularly rare, the species diversity (7 spp.) is not high. Especially noteworthy is the absence of the Black-faced Cotinga (*Conioptilon mcilhennyi*), which has been recorded as near as the north bank of the Río Madre de Dios, some 80 km NNE of the Reserve (Davis et al. 1991).

Southwestern Amazonia is known, not only for its high species diversity, but as an area with a high degree of endemism as well. Approximately two-thirds of the bird species endemic to the lowlands of southwestern Amazonia are found at the Explorer's Inn Reserve. Prominent among these are several species of especially small distribution, such as *Nannopsittaca dachilleae*, *Brachygalba albogularis*, *Malacoptila semicincta*, *Eubucco tucinkae*, *Myrmeciza goeldii*, *Percnostola lophotes*, *Cercomacra manu*, and *Poecilotriccus albifacies*. The distributions of these species are wholly centered on Madre de Dios and immediately adjacent areas of southern Perú and northern Bolivia.

Much of what has now become "common knowledge" regarding the vocalizations, habitat preferences and behaviors of Amazonian birds was totally unknown when Explorer's Inn was first visited. It is instructive to note that Appendix 3 includes seven species of birds—often cited as among the world's best-known organisms—that have been described as new to science within the past 30 years, in one case as recently as 1991.

Herpetofauna of Southeastern Perú (L. Rodríguez)

Species richness of anurans seems to vary at a rate of 10% both between localities and between habitat types (Rodríguez, in press). In the Madre de Dios area, several localities have been sufficiently sampled to provide reliable data on their species richness. Explorer's Inn is reported to have 70 species of anurans and 74 of reptiles (McDiarmid and Cocroft, unpub. data). Cuzco Amazónico has recorded 145 amphibians and reptiles, 65 of those are frogs (Duellman and Salas 1991). Pakitza on the Río Manu has 62 known species of amphibians and 51 reptiles (Morales and McDiarmid, unpub. data), while Cocha Cashu, also on the Río Manu, is known to have 82 amphibian species (80 frogs) and 64 reptiles (Rodríguez and Cadle, 1990). Combining all anuran species reported from all localities gives a rough estimate of the total diversity in Madre de Dios as follows:

Explorer's Inn Reserve	=	70
Ccolpa de Guacamayos	=	77
Cerros del Távara	=	83
Cuzco Amazónico	=	90
Pakitza (Manu)	=	100
Cocha Cashu (Manu)	=	107 spp.

Of these 107 species known from the Madre de Dios area, six (5.6% of the total) were recorded for the first time during the RAP expedition in the Tambopata-Candamo and a total of 13 spp. (15.6%) are new reports for the Tambopata-Candamo Reserved Zone. These data show variation in species composition among sites and the necessity of preserving as many habitat types as possible in order to maintain total diversity.

Together with the considerations given above about the species richness at the base of the Andes, we should also consider the geographical location of the Tambopata-Candamo area with respect to the Río Madre de Dios. The Madre de Dios river seems to be a physiographic boundary for some species of frogs such as *Edalorhina perezi* and *Scarthyla ostinodactyla*, which are common species at Cuzco Amazónico, Pakitza and Cocha Cashu,

Approximately two-thirds of the bird species endemic to the lowlands of southwestern Amazonia are found at the Explorer's Inn Reserve.

but absent from Explorer's Inn and localities on the right bank of the Madre de Dios. For these species, with ranges extending at least to the Iquitos region, the river may constitute their southern limit of distribution. On the other hand, species such as *Eleutherodactylus* cf. *cruralis*, *Epipedobates* sp. nov. and *Hyla callipleura*, seem to belong to a southern subset. Altitudinally, the herpetofauna from the Andean slopes (Távara-Guacamayo), are low montane elements, different from lowland elements. Although Cadle and Patton (1988) noted that an altitudinal segregation is more obvious among small mammals than in the herpetofauna of this area, a finer differentiation on the lower slopes (below 1500 m) should be carefully considered as more data becomes available.

It is likely then, that the Tambopata-Candamo area is of primary interest because it includes not only distributional boundaries for Andean and non-Andean species, but also central and southern Amazonian faunas. Furthermore, together with Manu National Park on the western bank of the Río Madre de Dios, this area will ensure habitat preservation for natural dispersion of populations. The phenomenon, largely depends on dispersing abilities of the species. In frogs, judging from their high fidelity to reproductive sites, these processes must be slow.

Butterflies of the Explorer's Inn Reserve (G. Lamas)

The butterfly fauna of the Explorer's Inn Reserve has been surveyed sporadically by myself and other lepidopterists since 1979. Appendix 12 lists 1,234 species recorded so far from an area of about 2.0 km² inside the Explorer's Inn Reserve. This list may increase somewhat when several species complexes are resolved, as sibling species may be included under a single name in some difficult genera. Most species are typically Amazonian-Guianan, but subspecies belong mostly to the "Inambari" endemism center. This represents the second richest documented butterfly community in the world, the first one being at Pakitza, Manu National Park, also in Madre de Dios, Perú, where over 1,300 species have been recorded. There are obvious affinities with the butterfly faunas found at Explorer's Inn and Pakitza. However, only 60% of the total number of species (1,591) found in Tambopata and Pakitza occur in both places. This preliminary result of a comparison of both faunas is quite surprising, as the areas are generally similar in forest cover and physical parameters, and lie at almost the same altitude (Pakitza is at 400 m). Some 3,600 species of butterflies are known for Perú; thus, at least one-third of the butterfly fauna of the country has been recorded at two small sites in Madre de Dios (Tambopata and Pakitza), highlighting the extraordinary diversity of the area.

After 13 years of rather sporadic surveying in the Explorer's Inn Reserve, I estimate that about 80% of its butterfly species have been recorded to date; however, every visit to the place produces new records. This was proven again during the RAP visit, when ten effective surveying days (June 3-11, 1992) were spent at the Reserve. In this time at least seven new species were recorded for the area, despite the fact that butterfly populations were very low as a result of unfavorable climatic conditions (two days before the initiation of the survey, the area suffered the effects of a severe cold front, which seriously depleted butterfly populations there).

Some 80 surveying hours were spent walking along several trails inside the forest, and at the banks of the Ríos Tambopata and La Torre. In all, 407 specimens were collected, representing less than 200 species, as a special effort was made to collect only rare or

... at least one-third of the butterfly fauna of the country has been recorded at two small sites in Madre de Dios ....

hard-to-identify species. Lycaenidae, Papilionidae and Dismorphiinae were almost absent, as is commonly the case during the height of the dry season, while Riodinidae, Ithomiinae and Satyrinae were scarce. As usual, hesperiids were the dominant group in diversity, but with very low population densities. Mud-puddling aggregations at river-banks were comparatively small, and consisted mostly of limenitidines and a few pierids.

RÍO HEATH AND PAMPAS DEL HEATH

The Río Heath forms part of the border between Perú and Bolivia. It also divides a cluster of seasonally-inundated savannas (known locally as "pampas"), most of which are on the Bolivian side. The Peruvian pampas have been recognized for some time (e.g., Hofmann et al. 1976) as a unique habitat within Peruvian territorial limits, and home to many species known nowhere else in the country. It consequently has been designated as a National Sanctuary, and with help of outside funding (mainly by the Nature Conservancy), now has an infrastructure of buildings and people to insure its protection.

Bolivia, however, is a country with great expanses of savanna of various kinds throughout the upper Amazon drainage and has yet to fully appreciate the importance of the Pampas del Heath and its status as one of the very few pampas in the country that has had almost no grazing by cattle. On cursory inspection it appears true that much of the fauna and flora of these pampas is also to be found elsewhere in northeastern Bolivia—though not yet actively protected anywhere.

Except for the military posts and lumbering operations at the mouth of the Río Heath, there is little colonization of the river valley. On the Peruvian side, above the last few meanders of the river are only a few indigenous settlements of the Ese'eja and three stations of the Pampas del Heath National Sanctuary. On the Bolivian side, while there is some mahogany extraction going on, the only significant incursion is the Medina family cattle ranch.

The team of Foster, Emmons, Lamas, Romo, Alban Castillo, and Wust, along with personnel from the Heath Sanctuary visited the Río Heath from June 12 to 20 1992, spending two days on the river (one night at Puesto San Antonio, and brief stops at Finca Medina and ccolpas along the river; 3 days working on the pampas at Juliaca, and a one-day trip upriver to the entrance to the Bolivian pampas (pampas de Mojos). The birds of this area were first surveyed by a party from the Louisiana State University Museum of Natural Science in June-July 1977 (Graham et al. 1980). Specimens collected on that trip are deposited at LSU and MUSM. Parker and Wust made additional observations on the birds here (Bolivian side) in August 1988 (Parker et al. 1991). Additional information on plants came from earlier visits to the Heath by Gentry and Nuñez in 1988.

Río Heath and Its Vegetation
(R. Foster and J. Alban Castillo)

The Río Heath appears to be significantly different from all the other rivers in Perú. It has much more in common with many of the rivers in northern Bolivia coming out of the lower ridges of the Andes (i.e., not the Río Beni), and with rivers draining the Brazilian Shield in eastern Bolivia.

Floodplain

The active floodplain of the river is very narrow. Meander development is very slow. There is no significant formation of beaches until halfway up river to the Juliaca entrance to the Peruvian pampas. The stands of *Ficus insipida*, characteristic of upper Amazon meander suc-

The Peruvian pampas have been recognized ... as a unique habitat ... and home to many species known nowhere else in the country.

... *sartenjales*

have an un-

usual thick

scrubby veg-

etation with

a high density

of lianas and

an open

canopy ....

cession, is poorly developed here. Even when frequent, it grows slowly, has few leaves, is covered with vines and parasitic Loranthaceae, and has no understory of large monocotyledonous herbs. There is much more *Acacia loretensis* than is customary, and these are loaded with parasitic Loranthaceae. *Calycophyllum spruceanum* is more abundant than usual. *Iriartea deltoidea* and *Alchornea castaniifolia* (Euphorbiaceae), though common on the lower part of the river, seem to disappear in the upper part. A "weeping" bamboo not seen on other Peruvian rivers was occasional on the bank, along with the more common *Guadua weberbaueri*.

Sartenjales

On either side of the active floodplain are extensive raised clay flats known as *sartenjales* or *shebonales* (the former refers to the flatness, the later to the characteristic palm, the shebon, which in this instance refers to *Attalea butyracea*). From what we could ascertain, these flats are not seasonally inundated by the river. But they are so poorly drained as to create a shallowly inundated forest from the rains of the wet season.

The sartenjales have an unusual thick scrubby vegetation with a high density of lianas and an open canopy of trees usually not much more than 20 m tall. Underneath is the distinctive hillock/channel surface, with the plants mostly occupying the hillocks and the water and terrestrial animals moving through the channels between them.

Occasionally the river meanders cut into one of these sartenjales exposing the substrate. These exposures are the small ccolpas on the Río Heath that attract so many parrots and other animals. The sartenjales sit on a deep (several meters), fine, heavy clay with a distinctive white and pinkish-red mottling. This in turn appears to rest on top of a layer of white

sand, but so far down that roots do not come close to reaching it. The abundance of lianas and of *Attalea butyracea* on these sartenjales suggests that this may be a fertile clay, such as montmorillonite, with a high cation exchange capacity rather than a more typical acidic kaolinitic clay. It may also explain greater concentrations of salts developing from the seasonal wetting and drying, and thus the attraction to animals. At this point it is all conjecture. These clays may be better at complexing toxic chemicals in the diet of animals (as suggested by C. Munn for the macaws). It remains to be seen whether this clay under these sartenjales is similar to that of the major ccolpas of the Río Tambopata and Río Manu.

The origin of the sartenjales and their hillock/channel surface is an interesting problem. Phillips (see Explorer's Inn section) has suggested that they are derived from *Mauritia* palm swamps that have gradually filled in. The decomposed bases of the *Mauritia* stems would thus form the basis for the raised tiny islands that become the hillocks and support the woody plants above the flood level. This hypothesis makes sense but may be too restricted. The hillock/channel surface is found commonly in many areas of Bolivia where *Mauritia* does not appear to be part of the ecosystem. I would suggest that any swamp forest, whether it be dominated by any palm species or other tree species, if it gradually fills with fine clay and becomes raised relative to the current floodplain will undergo the same change and develop a similar physiognomy.

This idea would predict that sartenjales might be very distinct from each other according to the parent material that formed the clay. Some of them (a more alkaline clay?) when eroded would be a source of ccolpas and would be thick with lianas, others, perhaps with a much more acid clay, would not form ccolpas and would be full of straight, small-crowned trees and shrubs, and with few lianas. Cursory observation of sartenjales throughout eastern Bolivia suggest that such a range exists. The importance of making such distinctions may be quickly obvious if we can accurately predict where ccolpas can be found throughout the Amazon Basin.

Another hypothesis to explain the sartenjales and the hillock/channel phenomenon is that it is derived from former pampas and their multitude of small hills formed by ant colonies and/or termites. These form tiny islands in the pampa above the level of inundation and on which colonization by woody plants usually has a head start when the pampa reverts to forest. Although the hillocks in most of the sartenjales seem too small and dense to be derived from the larger, more-dispersed ant/termite mounds, it may be with more exploration of former pampas areas these mounds will be found to be important. Given the abundance of palm swamps and flats within the pampas as well as areas of mounds, one might expect a mosaic of microtopography and vegetation composition in the pampas that have reverted to forest.

Whatever the origin of mounds or hillocks, the larger terrestrial vertebrates usually walk around them. This phenomenon, by trampling and erosion, accentuates the formation of channels. It can be seen in action from all the animal trails such as of peccaries, tapir, and deer in both swamp forests and in the pampa.

To gain access to the northern end of the larger Peruvian pampas from a place called Paujil, one has to ascend a high river bank and, further in, a steep slope up to the terrace of the pampas. But upriver at the Juliaca station, the rise in elevation from the river bank to the central pampas region is gradual and barely noticeable. Further upriver, the access to the Bolivian pampas is also short and the rise in elevation very minor. This apparently reflects a gradual merging—as one approaches the Andes—of the sartenjal terrace of the river and the terrace of the pampas. Based on the LANDSAT image and over-flights between the Heath and Alto Madidi on the Bolivian side, this broad, flat area east of the Río Tambopata and just north of the foothills all the way down to the Río Beni and beyond, is a low open forest dominated by palms. It remains to be seen whether this has the hillock/channel surface typical of sartenjales, or whether it is some intermediate vegetation type subject to annual or subannual river flooding and sedimentation.

Pampas del Heath and their Vegetation (R. Foster, J. Alban Castillo)

There are two "active" pampas on the Peruvian side of the border and both are small. They were apparently continuous at one time (see below), forming a pampa closer in size to those on the Bolivian side. The smaller of the two, and least visited, is to the north and appears on the satellite image to be rapidly converting to shrub thickets and forest. We studied the larger pampa, which is accessible by a 30-40 minute trail walk from the Juliaca camp on the river.

Recently-burned Pampa

No matter where you are in the Peruvian Pampas del Heath, you can always see the forest edge. It is much smaller than the pampas on the Bolivian side and probably of less significance biologically. Getting through it is however very difficult because of the very thick grass that tangles the feet and the frequent wet depressions for much of the year.

The surface water of the pampa drains to either side, collecting in the lowest depressions along the forest edge, usually an aguaje stand. The runoff forms crystalline streams as it enters the forest. Not far into the forest (sometimes only a few meters, sometimes a hundred), the stream drops off the terrace of the pampas in a waterfall 2-3 m high. The waterfall exposes the substrate, which consists of about 1 m of black (burned) organic material overlaying a pink-white blotchy clay similar to (if not the same as) that seen at the ccolpas along the river.

The pampa is clearly burned at irregular intervals and not consistently in the same areas. From the pattern of recent burns on the forest edges, it appears that the fire usually burns from the southeast to the northwest.

Though parts of the pampa are flat and relatively dry, most of the pampa surface is of alternating low swales and humps. It is not clear whether this is the preserved topography of former alluvial action, or whether these are induced by ants/termites. Certainly the large mounds on the drier areas with high grass are active termite mounds now.

The density of shrubs and small trees out in the pampas seems related to the interval between big fires. The best development of trees is usually on the large termite mounds or in permanently wet swales.

There are dozens of grass species in the pampa and our specimens are awaiting identification by Dr. Oscar Tovar, grass specialist at the Museo de Historia Natural in Lima. Among the common dicotyledonous herbs are *Chamaecrista* spp. (Leguminosae-Caesalp.), *Cuphea repens* (Lythraceae), *Desmoscelis villosa* (Melastomataceae), *Sipania hispida* (Rubiaceae), *Tephrosia sinapou* (Leguminosae-Papil.), and there are dozens of others. None of the herb species seem to be found throughout the pampa.

The family Melastomataceae makes up a large proportion of the shrubs. The two most common and most widespread are *Macairea thyrsiflora*—especially conspicuous in season with its copious production of bright pink flowers—and *Graffenrieda weddellii*, with vertically held folded leaves distinctively white on the underside. Other common melastomes

are *Bellucia grossularioides*, *Clidemia capitellata*, and several *Miconia*, such as *M. albicans*, *M. mattogrossensis*, *M. rufescens*, and *M. tiliifolia*. There are several shrubby Rubiaceae such as *Alibertia* spp., Myrtaceae such as *Myrcia guinanensis*, a very common Polygalaceae, *Bredemeyera lucida*, and dozens of other less common species. The most common trees on the mounds are *Graffenrieda limbata*, *Matayba guianensis* (Sapindaceae), *Virola sebifera* (Myristicaceae), *Hirtella* sp. (Chrysobalanaceae), *Xylopia* sp. (Annonaceae), *Myrcia paivae* (Myrtaceae), *Himatanthus sucuuba* (Apocynaceae) and various Rubiaceae, especially *Remijia firmula* and *Ladenbergia graciliflora*. At least four (and probably several more) of the species in our collections from the pampas are new records for Perú.

A striking feature of the floristic composition of the pampa is the patchiness of most of the plant species, whether grasses, shrubs, or tree. Each mound, and each region of grass seems to have a unique collection of species. A detailed study may reveal how much of this heterogeneity is a result of different periods of inundation, different fire frequency, or just the probabilities of dispersal and colonization.

Forest edge and forest islands often have an unusual, uniform appearance on the outside because of a few small-leaved pampas trees, such as *Hirtella* sp., *Xylopia* sp., and *Myrcia paivae*. Behind this front wall are sometimes dense stands of what looks like a giant banana, *Phenakospermum* sp. (Strelitziaceae). These giant herbs are monocarpic, producing one enormous terminal inflorescence of bat-pollinated flowers and bird-dispersed seed before dying.

Islands of forest within the pampas usually represent successional stages of pampas regeneration in areas that have not recently been hit by fire for some reason, usually because of a water barrier. Probably they would burn in the rare, exceptionally dry years when some of the water barriers dry up. Within these islands is usually a combination of wet channels that are also tapir trails, and high old ant/termite mounds. When the islands are less than about 10 m diameter, the forest is composed mainly of trees that are characteristic of the pampas termite mounds. When larger than 10 m, many "true" forest species start to enter the islands (including the conspicuous palms *Jessenia* and *Euterpe*, and the banana-relative *Phenakospermum*), while the mound species become mostly confined to the edges. The mound species are presumably more resistant to fire.

The *Mauritia flexuosa* (aguaje) palms are mostly in dense patches of the wet depressions in the center of the pampa and along some edges. Occasional isolated individuals are scattered in the wetter grassy areas. Most puzzling is what seems to be straight lines of aguajes running through or across the middle of the pampas. Perhaps these have colonized deeply rutted and flooded animal tracks, or mark some earlier channel digging by indigenous people.

In parts of the pampa are permanent pools of water (espejos de agua) several meters across. The shallower parts of these pools have many plants not seen elsewhere in the pampa such as certain sedges, tiny white *Utricularia* species, and floating and submerged aquatics.

Tapir and marsh deer trails are visible throughout the pampa, at ground level and from the air. The most heavily grazed areas are on the west side, which is thick with the ruts of animal trails. This may reflect the animals having learned that hunters and other visitors always enter the pampa from the river side on the east. The west side is also mostly wetter pampa compared to the east.

Clearly, if there is to be any hope of maintaining the Pampas del Heath as a sa-vanna, deliberate burning is the most important element. To maintain the diversity of the habitat and the diversity of species it should not all be burned at once, nor every year. A rotating system of burning for different sections of the pampa—following the guidance (timing, wind-direction, etc.) of the people who have lived and hunted in the area for many years must be implemented. It would be advisable to enlarge the area of open savanna by more intensive burning of those areas that were most recently savanna (see below), such as the large isolated area of young regenerating pampa to the north of the current pampa.

Regenerating Pampa

Downriver at Paujil we entered the north end of the pampa where the forest is growing back. It is here that we saw what has apparently been happening to much of the former pampas over the last decades. According to our informant, Darío Cruz, he and one of the indigenous leaders had set fire to this area 10 years earlier to maintain the grassland, but it had not been burned since then. The central area of the Peruvian pampa has been burned more than once since then, but there are many small pockets along the edge that have escaped these fires and evidence regeneration of woody plants and the disappearance of the grass similar to this larger area.

In the post-fire succession, shrubs and small trees were spreading out from the ant/termite mounds and from the forest edge. This has produced a vegetation that consists of clumps of shrubby thickets with only minor grassy openings and animal trails between and among them. With the increased shade and root competition from the woody plants, the grasses remaining were not developing reproductive parts, and most of the other herbs typical of the open pampa were missing. *Selaginella* and *Lycopodium* dominated much of the ground cover. A few sedges were in

Tapir and marsh deer trails are visible throughout the pampa ....

> **The better-drained the regenerating forest is, the more mixed the species composition.**

fertile condition, and there were some small *Cuphea* (Lythraceae) flowering. In the drier areas with more mound development the common small trees were much the same as those found on the mounds in the central burned pampas. In the wetter areas, the common trees, 4-5 m tall, were *Qualea wittrockii* (Vochysiaceae), *Licania* sp. (Chrysobalanaceae), and two Rubiaceae (*Pagamea guianensis* and *Ladenbergia graciliflora*), also sometimes mixed with *Mauritia*. Epiphytic ant-gardens were common on the trees and usually included *Codonanthe* (Gesneriaceae), *Clusia* (Guttiferae), three orchid species, one bromeliad, and *Epiphyllum* (Cactaceae).

Older successional forest in the wetter areas is usually dominated by large trees of *Qualea wittrockii*, sometimes forming nearly solid stands more than a kilometer across, and sometimes mixed with palms and a few other tree species. The better-drained the regenerating forest is, the more mixed the species composition. There is much to be explored and learned about the composition and pattern of post-fire pampas succession.

From the satellite image it is obvious that much of the high terraces to the west of these active pampas, between the Río Heath and the Río La Torre, are forests that are in some stage of regeneration from former pampas. This is to say that much of the forest area of the lower Tambopata-Candamo Reserved Zone is apparently high terrace of an impermeable clay and was formerly a burned, and seasonally-flooded savanna. It appears to be mostly the northern extremes of these high terraces—toward the Río Madre de Dios— that have the sandy, well-drained soil supporting the castañas (Brazil nut trees), such as the sandy high-terrace at the eastern end of the Explorer's Inn property. But not far south and east of Explorer's Inn is what appears to be old regenerating forest of a former pampa.

As a first hypothesis, I would suggest that the former northwestern pampas were maintained due to the deliberate burning by the indigenous people, presumably to maintain game animals. The abandonment of the area by the indigenous people probably coincides with the arrival of many outsiders to the Puerto Maldonado area during the rubber boom of 100+ years ago—with all the attendant butchery, enslavement, and disease. These pampas might be relics of a much earlier time when the climate was much drier, and they have been maintained by human initiated burns since the climate changed for the wetter. Alternatively, they might have once been forested and converted to grassland by humans to extend the range of the game animals from the south. Perhaps some modern archeological/paleontological techniques will eventually reveal this history.

The Río Heath area may have been an indigenous refuge for many decades since the rubber boom. However, the current Ese'eja settlement on the lower Heath appears to be a result of recent immigration from the Madidi-Ixiamas area in Bolivia. There is much recent history in the area that needs clarification, and a thorough study undertaken soon would find many people still alive who know what has been going on, at least for the last 50 years.

Birds of the Pampas del Heath
(T. Parker, T. Schulenberg and W. Wust)

These pampas have high bird species diversity, especially so considering that they represent the very westernmost extension of pampas habitat (Appendix 4). Fourteen species are known in Perú only from the Pampas del Heath, and several additional species of the Pampas are recorded from only two or three other sites within Perú. As one would expect, the species of the Peruvian pampas are similar to those of more extensive pampas farther east in Bolivia; most of them are more or less widely distributed across central South America in grassland habitats. Although lacking in bird species of global rarity, the Pampas del Heath may be an important refuge for large mammals (Parker and Bailey 1991). The near pristine conditions of the Pampas del Heath are further exemplified by the impressive abundance of macaws.

Mammals of the Río Heath and Peruvian Pampas (L.H. Emmons, C. Ascorra, and M. Romo)

The Peruvian Pampas del Heath is the northwesternmost fragment of a large chain of pampas that extends east and south through Bolivia and Brazil at the southern fringe of the Amazonian rainforest, through areas of dry and wet forests, and ultimately to pantanal, cerrado and chaco habitats. The mammal faunas of the northern complex of wet and dry grasslands, such as those of the Beni region, lack the high mammalian diversity of the cerrado (Redford and Fonseca 1986). We can speculate that this may result from a great instability of resource conditions, due to frequent flooding and burning, and perhaps to a relatively young geologic age or impoverished soils. Almost all of these pampas are now used to raise cattle, and frequented by ganaderos who hunt game for their subsistence. Where domestic livestock occur, the pampas are severely to partially altered by their activities, as well as those of the human tenants. The Pampas del Heath have miraculously escaped conversion to pasture and stand as one of the last examples of unaltered native flora and fauna of northern, forest-edge grasslands, except for probable pre-Columbian Amerindian use. These grasslands are relicts of a former climatic regime that was drier than the present, and ironically they are now only preserved from invasion by forest by the opposing forces of fire and standing water.

The mammal fauna of the Pampas del Heath has not been surveyed on either the Peruvian or Bolivian side of the Río Heath, but larger mammals were collected on an earlier expedition, including marsh deer, maned wolf and giant anteater (Hofmann et al. 1976). In 1977, members of Louisiana State University (LSU) expeditions collected some mammals in the Pampas in the course of studies of birds, and these records are published for the first time here. This is thus the first complete documentation of the mammal fauna known in the pampas and surrounding park. The open habitats of the Pampas del Heath are surrounded by closed evergreen forest, and such forest comprises a large part of the National Sanctuary. On this expedition we surveyed mammals both in the pampas and forest of the Sanctuary and along the banks of the Río Heath.

We recorded a total of 68 species of mammals in the region, 15 of these in the open pampa, and 52 in the closed-canopy forest or on the riverbank (Appendix 7). Another six species were found by earlier expeditions (Hofmann et al. 1976; LSU), bringing the total recorded for the park to 74 species. The forest mammals of the Río Heath basin are a subset of the mammals already known from lowland Madre de Dios (Ascorra et al. 1991, Janson and Emmons 1990, Woodman et al. 1991, Emmons and Romo, this report). However, two records by members of our groups are particularly noteworthy.

Ascorra et al. sighted a pacarana (*Dinomys branickii*) on a bank of the Río Heath. This species of giant rodent seems extremely scarce in most of its range (except perhaps in a part of Acre; K. Redford, pers comm.) and is rarely reported. It is mainly found on the eastern Andean slopes, and only enters the lowlands in southern Perú, northern Bolivia, and adjacent Brazil. Local people stated that pacaranas were regularly found in

The Pampas del Heath have miraculously escaped conversion to pasture and stand as one of the last examples of unaltered native flora and fauna of northern, forest-edge grasslands ....

The gallery forest ... included a number of species typical of the surrounding rain forest, such as monkeys, tamarins, bats, rodents and opossums ....

the area. Pacaranas are hunted for meat and we feel that the species is rarer and potentially much more threatened than many species that are considered endangered, but that are in fact extraordinarily common and widespread, such as ocelots and jaguars.

Romo and Lamas saw a short-eared dog (*Atelocynus microtis*) in the forest a short distance from the Pampas at Refugio Juliaca. This species, also extremely rare throughout its range, was likewise reported to be quite common on the Río Heath by a lifetime resident of the region (Darío Cruz Vani). We also heard reports of its being common in Pando (where we also probably sighted it). Coupled with our recent encounter of the species on the upper Río Madidi (Parker and Bailey 1991), there is evidence that the Madre de Dios and northern La Paz drainages hold a major population of this species (perhaps the only large one).

In the pampas, members of both groups saw single individuals of marsh deer (*Blastocerus dichotomus*), a male and a female. We also collected three grassland rodents, *Cavia aperea*, *Oryzomys buccinatus* and *Bolomys lasiurus*; the latter two of which are new records for Perú. The remaining species that we encountered within the grassland were forest species of wide occurence. Maned wolves (*Chrysocyon brachyurus*, CITES Appendix II) were previously found in the Peruvian pampas (Hofmann et al. 1979; Darío Cruz V., pers. comm.), but we found no evidence of their presence, despite a search for spoor. However, subsequent to our expedition, park guards collected some maned wolf feces from somewhere in the Sanctuary (V. Pacheco, pers. comm.).

The gallery forest (Aguas Claras Camp) included a number of species typical of the surrounding rain forest, such as monkeys, tamarins, bats, rodents and opossums (Appendix 7). The edges of the gallery forest and

newly-formed woodland that has covered former pampa has a groundcover of almost pure *Selaginella* that has replaced the grasses and sedges. This vegetation seems devoid of small and large mammals; no mammals were captured in traps in it, and the only large mammal tracks in a large wooded area (locality 11) were those of tapirs (which travel everywhere). Marsh deer apparently avoided this cerrado-like vegetation. Preliminary results from analysis of bat feces (Ascorra, in prep.) suggest that phyllostomid bats are the chief dispersers of several woody plants, such as *Vismia* sp., into the pampas. Bats are thus agents for regeneration of the forest and closing over of the grasslands.

Our expedition was too short to exhaustively survey the pampas for mammals, and a few more grassland or cerrado species may be found there. One other savanna species is said to be found in the pampas—the seven-banded armadillo (*Dasypus septemcinctus*, Darío Cruz V., pers. comm.). This species has not been recorded in Perú, but is known from the Beni savannas in Bolivia. Future expeditions should try to verify this report.

We can therefore confirm the presence of four mammals that are strictly pampa species in Pampas del Heath National Sanctuary, and a fifth has been reconfirmed. Two others need to be verified. Of the five, four occur nowhere else in Perú (of the seven, six). The fauna is thus small, but significant. The most globally important mammals in the pampas are the marsh deer (CITES Appendix I). This large, elegant deer has been decimated by ranchers and cattle diseases (probably brucellosis) over most of its former range (e.g., Schaller and Vasconcelos 1978). The deer in the Pampas del Heath are one of the few populations not living among livestock, although the few cattle in the Bolivian pampas may expose the entire population of this widely-traveling deer to disease.

Marsh deer and maned wolves are both large species that live on big home ranges. The two species use different types of habitat, with the deer specialized for wet grasslands, and the maned wolves for dry ones, where they feed on rodents, armadillos, and fruit. The densest population of marsh deer reported was one per 3.8 km² (Schaller and Vasconcelos 1978). Hofmann et al. (1976) estimated from casual observations that from 1970-72, populations of marsh deer in the pampas were about 0.5-0.7/km², and those of maned wolf about 0.6/km² (ten times higher than reported elsewhere, Deitz 1984). They stated that in those years maned wolves were common and they saw many. We saw no evidence of them and suspect that the species is now greatly reduced in the largest of the Peruvian pampas. Only parts of the pampas are suitable for each species. The pampas may recently have shrunk to below the size needed for a breeding population of maned wolves (but there are many other possible causes for reduced populations). It risks to do so for both maned wolves and marsh deer unless carefully managed.

Herpetofauna of the Pampas del Heath (J. Icochea)

This opportunity to survey within the National Sanctuary has allowed our knowledge of the herpetofauna to be greatly augmented. When the proposal to establish the Pampas del Heath National Sanctuary was made (Herrera and Pulido 1982), the faunal data used to support it principally concerned mammals and birds. Only two species of reptiles were considered, white caiman (*Caiman crocodilus*) and the river turtle, "taricaya" (*Podocnemis unifilis*), which is understandable since they are the most visible elements of the herpetofauna. With this quick survey we are able to report 28 species of amphibians

and 17 reptiles (Appendix 9).

The amphibians are all anurans, the majority belonging to the families Hylidae and Leptodactylidae. Three species were found in three of the four principal areas collected: *Bufo marinus*, *Osteocephalus* sp. (red toes), *Eleutherodactylus fenestratus*, and *Leptodactylus wagneri*. Almost half of the species prefer terrestrial environments and more than half of those are arboreal. Only *Pipa pipa* is exclusively aquatic. The majority of the species were found in the wet forest next to the river. On the pampa proper where grass predominates, only one species was found, *Hyla* sp. 1, about 50 m from the edge of the grassland. In the creek that begins on the same pampa (headwater of Quebrada Tapir), in addition to the species just mentioned, two others were also captured, *Leptodactylus* sp. 1 and *Adenomera* sp. 1.

Reptiles recorded include two turtles, seven lizards, six snakes, and two caimans. Eleven of these species were collected and six were merely observed. All of the species were previously known from Amazonian Perú. *Ameiva ameiva* was the only reptile observed on the open pampa. An approximately 6 m long anaconda (*Eunectes murinus*) was seen in the quebrada coming out of the pampa (Q. Tapir) and we captured a juvenile specimen of the caiman *Paleosuchus palpebrosus*. On June 5th, I counted specimens of the river turtle (*Podocnemis unifilis*) on a section of the Río Heath between Refugio Juliaca and Refugio Picoplancha (ca. 10 km), from 12:00 to 14:16 h, with the temperature approximately 32° C. Two hundred sixty-nine individuals were counted basking on exposed tree trunks, a density of 26.9 turtles/km of river. Five specimens of *Caiman crocodilus* were seen on the banks of the river.

The number of species collected is considered reasonable given the short duration of the trip, the fact it was the dry season, and several days of cool weather that diminished the activity of amphibians. The species composition of the collection obtained in the Pampas del Heath can be compared to four other sites inventoried in Madre de Dios, Cocha Cashu (Rodríguez and Cadle 1990), Cuzco Amazónico (Duellman and Salas 1991), Explorer's Inn Reserve (= Tambopata; McDiarmid and Cocroft, unpub.), and Pakitza (McDiarmid and Morales, unpub.). Almost two-thirds of the species are shared with those four localities, but with a few species more in common with Tambopata and Cuzco Amazónico than with Pakitza and Cocha Cashu.

Given that the basin of the Río Heath includes similar habitats on both sides of the Perú-Bolivia border, there are probably many species in common with Bolivia. A preliminary inventory of the amphibians of Bolivia (De La Riva 1990) allows us to see that there are 14 species in common with the species already identified from the Pampas del Heath. These species are found in the ecoregion referred to as the "Llanura Tropical Húmeda" (Humid Tropical Plain), within which he includes the Bolivian pampas. Also, in comparing the few lizards collected with a preliminary list for Bolivia (Fugler 1989), five of the six species are shared with Bolivia.

I do not think that this area will be found to be as species rich as the other areas in Madre de Dios already mentioned, but the study of species unique to the Pampas del Heath merits additional attention. Additional sampling over several years will be needed in order to approach a complete inventory of the herpetofauna.

With respect to studies of the biology and ecology of a particular species, populations of river turtles (*Podocnemis unifilis*) in the Río Heath appear favorable for such purposes, especially considering the use made of this resource by the native inhabitants of the area.

Fish Fauna of the Pampas del Heath (H. Ortega)

In Amazonian Perú, we know the general ichthyofaunal composition of some of the large and medium-sized rivers of the lowland forest (Ucayali, Marañón, and Madre de Dios). Obvious advances in our knowledge include the recent anotated lists of the continental fishes of Perú (Ortega and Vari 1986, Ortega 1991), and continuing growth in the ichthyological collection of the Museo de Historia Natural (MUSM). On the other hand, there are numerous drainage basins and subbasins that are unknown with respect to their ichthyofauna, as is the case with the Río Heath, an affluent of the Río Madre de Dios.

Between May and June of 1992, we conducted a first, rapid evaluation of part of the Río Heath basin. Our results indicate a high diversity of fish species. This information will serve many whose interests are related to the management of the ecosystem and also provides more information for the development of specific resource conservation programs, especially aquatic resources.

The Pampas del Heath National Sanctuary is located in the Province of Tambopata, Department of Madre de Dios, between the Heath and Tambopata Rivers. The Sanctuary measures 102,109 ha. In the inventory of icthyology, the area of study included six bodies of water pertaining to the Heath River basin between the San Antonio gaurdpost and Refugio Juliaca. These were varying habitats (river, ravines, lagoons, streams, and seasonal pools)

Approximately 2400 specimens between 12 and 250 mm total length were collected. The results indicate the presence of 95 fish species in the aquatic habitats of the National Sanctuary. These species represent 74 genera and 23 families. Appendix 10 contains a summary by family and collecting station; Appendix 11 includes a systematic list of the species and the provenance of each species by collecting station. Around 10% of the specimens remain to be identified.

The species composition recorded in almost three weeks of field work in six bodies of water in the Pampas del Heath National Sanctuary demonstrates a high and interestingly diverse ichthyofauna. In comparison, 45 total days in the field between 1982 and 1991, at the Explorer's Inn Reserve, lead to a list of 120 species (Chang, 1991), and 120 days of field work between 1987 and 1991 in Parque Nacional Manu has recorded 175 species (Ortega, in prep.). Considering the effort and techniques used in the other studies, it is very likely that with two repetitions at other times and additional samples in adjacent habitats, e.g., Río Palma Real and similar affluents, the number of species could be doubled.

In overall composition, Characiformes and Siluriformes predominate, with eight and seven families respectively, followed by Gymnotiformes with three. With respect to the number of species per family, Characidae is the largest family with 42 species, followed by Pimelodidae (8), Loricariidae (5), and Curimatidae (5).

With respect to sampling station, the Río Heath has the greatest number of species (58), which reflects a high diversity in the river itself and in the mouths of the creeks where there occurs a mixing of waters. Picoplancha has 42 species and Quebrada San Antonio has 34. Environments with few fish species include the seasonal pools at San Antonio with only four species of forms resistant to desiccation (Rivulidae), and Quebrada Shuyo (4), which borders part of the pampa and has crystalline water with a relatively low temperature. Here exists a short, efficient food chain consisting of three small insectivorous species and one larger, predatory fish. This type of community is very unique for a lowland forest site (269 m).

Range extensions for some species (e.g., *Carnegiella strigata* and *Nannostomus trifasciatus*) that were previously known from the immediate vicinity of Iquitos in Loreto is notable, especially because their value as ornamental species makes them more conspicuous.

Considering the real and potential value of the fishes, both for those species consumed (25%) and the ornamental fishes (44%), it would be worthwhile to direct more attention to both groups for studies of biology, ecology, and fishing. It would also be interesting to combine these research activities with visits by technicians to introduce improvements in some processing methods, like salt-drying. Together, these activities would benefit the local inhabitants, fishermen, and the people involved.

Butterflies of the Pampas del Heath (G. Lamas)

There are no previous reports of butterflies from the Pampas del Heath. Six effective surveying days were spent at Pampas del Heath (June 14-19), yielding 203 species records. Appendix 13, lists the species from the area, recorded along the trail leading from the base camp to the savanna, on the savanna itself, and at gallery and island forest within the savanna. Some species were observed (not collected) at the banks of the Río Heath, close to camp. Approximately 55 surveying hours (almost equally divided between savanna and gallery-island forest) were spent at the site, and 532 specimens were collected. Presence and relative abundances of the different taxonomic groups were similar to those observed earlier at Explorer's Inn.

The species records include a remarkable 29 new species for Perú, of which about five are undescribed. Eighteen of the 29 newly recorded species for Perú were found only in the savanna, or barely entering gallery or

... Pampas del Heath National Sanctuary demonstrates a high and interestingly diverse ichthyofauna.

island forest, and only two of the new records constitute truly forest species (*Alesa prema* and *Cariomothis* sp. n.). Given the very short survey period, performed during a relatively unfavorable season, it is not unreasonable to predict a further 50-100 possible new records for Perú from this area.

Particularly interesting were the species restricted to savanna, as many of them were known only from far-away places in Brazil, like *Philaethria pygmalion* (its previous southwesternmost locality was on the lower Madeira river), *Audre middletoni* (only recorded from Goiás), and *Copaeodes castanea* (only known from Paraná), or Argentina, as *Strymon cyanofusca*, only known from the type-locality in Chubut. Most forest species were common and had been seen on prior days at Tambopata (Explorer's Inn Reserve), but remarkable findings include three species not yet recorded from Tambopata, *Alesa prema*, *Cariomothis* sp. n., and *Saliana chiomara*. The gallery forest within the savanna proved particularly rich in rare or uncommon species of Hesperiidae, like *Chrysoplectrum perniciosus*, *Zera tetrastigma*, *Quadrus fanda*, *Gindanes brebisson*, *Vettius lucretius*, *Carystus hocus*, and *Panoquina bola*.

Obvious affinities of the savanna species are with the cerrado fauna of Brazil, and the chaco of Paraguay and Argentina, an affinity shared by many of the species found within the gallery forest. Forest species were obviously related to those found at Tambopata and other sites in southeastern Perú, and no surprising discoveries were made among them, except the new species of *Cariomothis*.

As in Tambopata, the greatest diversity of forest butterflies at Pampas del Heath should be found at the end of the dry season (September-November), and the same could happen with the savanna species. Notwithstanding this, I believe that most savanna species present in June (i.e., its "standing crop") were recorded. Naturally, a more extended survey, including different seasons of the year, must be undertaken to get a clearer picture of the Pampas butterfly diversity. Probably well in excess of 1,000 butterfly species occur in the Sanctuary, so this six day survey recorded less than 20% of its diversity.

Particularly interesting were the species restricted to savanna, as many of them were known only from far-away places in Brazil ... or Argentina ....

Literature Cited

Allen, J.A. 1900. On mammals collected in southeastern Peru by Mr. H. H. Keays, with descriptions of new species. Bull. Amer. Mus. Nat. Hist. 13:219-228.

Ascorra, C.F., D.E. Wilson and M. Romo. 1991. Lista anotada de los quirópteros del Parque Nacional Manu, Perú. Publ. Mus. Hist. Nat., Univ. Nac. Mayor de San Marcos, Ser. A Zool. 42:1-14.

Cadle, J.E. and J.L. Patton. 1988. Distribution patterns of some amphibians, reptiles, and mammals of the eastern Andean slope of southeastern Peru. Pp. 225-244. *In*: P.E. Vanzolini and W. R. Heyer (eds.). Proc. of a Workshop on Neotropical Distribution Patterns. Acad. Brasileira Cien., Rio de Janeiro.

Campbell, K.E., Jr., C.D. Frailey and J. Arellano L. 1985. The geology of the Rio Beni: Further evidence for Holocene flooding in Amazonia. Contrib. Sci. 346:1-8.

Chang, F. l99l. Ictiofauna de la Zona Reservada de Tambopata, Madre de Dios, Perú. Tesis para Bachiller en CC BB. Universidad Ricardo Palma.

Davis, T.J., C. Fox, L. Salinas, G. Ballon, and C. Arana. 1991. Annotated checklist of the birds of Cuzco Amazonico, Peru. Occ. Pap. Mus. Nat. Hist., Univ. Kansas 144:1-19.

De la Riva, I. 1990. Lista preliminar comentada de los anfibios de Bolivia con datos sobre su distribución. Boll. Mus. Reg. Sci. Nat. - Torino 8:261-319.

Dietz, J.M. 1984. Ecology and social organization of the maned wolf (*Chrysocyon brachyurus*). Smithsonian Contrib. Zool. 392.

Duellman, W.E. and A.W. Salas. 1991. Annotated checklist of the amphibians and reptiles of Cuzco Amazónico, Peru. Occas. Pap. Mus. Nat. Hist., Univ. Kansas 143:1-13.

Flores, G. and L.O. Rodríguez. MS. Two new species of *Eleutherodactylus* from Peru. Copeia, submitted.

Fugler, C.M. 1989. Lista preliminar de los saurios. Ecología en Bolivia 13:57-75.

Gentry, A.H. 1988a. Tree species richness of upper Amazonian forests. Proc. Nat. Acad. Sci., USA 85:156-159.

Gentry, A.H. 1988b. Changes in plant community diversity and floristic composition on environmental and geographical gradients. Ann. Missouri Bot. Gard. 75:1-50.

Gentry, A.H. 1989. Checklist of the plants, Zona Reservada de Tambopata, Perú. Missouri Botanical Garden, unpublished.

Graham, G.L., G.R. Graves, T.S. Schulenberg, and J.P. O'Neill. 1980. Seventeen bird species new to Peru from the Pampas de Heath. Auk 97:366-370.

Greenwood, P.H., D.E. Rosen, S.H. Weitzman, and G.S. Myers. 1966. Phyletic studies of teleostean fishes, with a provisional classification of living forms. Bull. Amer. Mus. Nat. Hist. 131:339-456.

Herrera F-D, E. and V. Pulido C. 1982. Informe para el establecimiento de una Unidad de Conservación Pampas del Heath. Informe No. 24-A-82-DGFF-DRFF, Ministerio de Agricultura y Alimentación.

Hofmann, R.K., C.F. Ponce del Prado and K.C. Otte. 1976. Registro de dos nuevos mamíferos para el Perú, *Odocoileus dichotomus* (Illiger-1811) y *Chrysocyon brachyurus* (Illiger-1811) con notas sobre su hábitat. Rev. Forest. Perú 6:61-81.

INRENA. 1994. Informe Tecnico Sobre la Propuesta para el Establecimiento del Parque Nacional Bahuaja-Sonene (Tambopata-Heath). Ministerio de Agricultura, Lima. 58 pp.

Janson, C.H. and L.H. Emmons. 1990. Ecological structure of the nonflying mammal community at Cocha Cashu Biological Station, Manu National Park, Peru. Pp. 314-338. *In*: A. Gentry (ed.) Four Neotropical Rainforests. Yale Univ. Press.

Kratter, A.W. In press. Natural history, population status, and conservation of the Rufous-fronted Antthrush (*Formicarius rufifrons*). Bird Conservation International.

Lowe-McConnell, R. 1987. Ecological Studies in Tropical Fish Communities. Cambridge Univ. Press. 371 pp.

Nelson, J.S. 1984. Fishes of the World, 2nd ed. Wiley-Interscience. 523 pp.

Ochoa, C. 1975. Potato collecting expeditions in Chile, Bolivia and Peru, and the genetic erosion of indigenous cultivars. pp. 167-173. *In:* Crop Genetic Resources for Today and Tomorrow — International Biological Program. Cambridge Univ. Press.

Ortega, H. l991. Adiciones y correcciones a la lista anotada de los peces continentales del Perú. Publ. Mus. Hist. Nat., Univ. Nac. Mayor de San Marcos, Ser. A Zool. 39:l-6.

Ortega, H. En preparación. Ictiofauna del Parque Nacional Manu, Madre de Dios, Perú. Simposio BIOLAT.

Ortega, H. and R.P. Vari.l986. Annotated checklist of the freshwater fishes of Perú. Smithsonian Contr. Zool. 437:l-25.

Osgood, W.H. 1944. Nine new South American rodents. Field Mus. Nat. Hist., Zool. Ser. 29:191-204.

Parker, T.A., III. 1982. Observations of some unusual rainforest and marsh birds in southeastern Peru. Wilson Bull. 94:477-493.

Parker, T.A., III. 1991. On the use of tape recorders in avifaunal surveys. Auk 108:443-444.

Parker, T.A., III and B. Bailey (eds). 1991. A biological assessment of the Alto Madidi region and adjacent areas of northwest Bolivia, May 18-June 15, 1990. Conservation International, RAP Work. Pap. 1:1-108.

Parker, T.A., III, A. Castillo U., M. Gell-Mann, and O. Rocha O. 1991. Records of new and unusual birds from northern Bolivia. Bull. Brit. Ornith. Club 111:120-138.

Phillips, O.L. 1992. *Ficus insipida* (Moraceae): etnobotánica y ecología de un antihelmíntico Amazónico. Rev. For. Perú 19:91-95.

Phillips, O.L. 1993a. The potential for harvesting fruits in tropical rainforest: new data from Amazonian Perú. Biodiv. and Conserv. 2:18-38.

Phillips, O.L. 1993b. Comparative Valuation of Tropical Forests in Amazonian Peru. Ph.D. thesis. Washington Univ., St. Louis, Missouri.

Phillips, O.L. and A.H. Gentry. 1993a. The useful plants of Tambopata, Perú. I: Statistical hypotheses tests with a new quantitative technique. Econ. Bot. 47:15-32.

Phillips, O.L. and A.H. Gentry. 1993b. The useful plants of Tambopata, Perú. II: Additional hypothesis testing in quantitative ethnobotany. Econ. Bot. 47:33-43.

Phillips, O.L. and A.H. Gentry. 1994. Increasing turnover through time in tropical forests. Science 263:954-958.

Phillips, O.L., P. Hall, A.H. Gentry, S.A. Sawyer and R. Vasquez. 1994a. Dynamics and species richness of tropical rain forests. Proc. Nat. Acad. Sci., USA 91:2805-2809.

Phillips, O.L., A.H. Gentry, C. Reynel, P. Wilkin, and C. Gálvez Durand B. 1994b. Quantitative ethnobotany and Amazonian conservation. Conserv. Biol. 8:225-248.

Pimm, S.L. and A.M. Sugden. 1994. Tropical diversity and global change. Science 263:933-934.

Redford, K.H, and G.A.B. da Fonseca. 1986. The role of gallery forests in the zoogeography of the Cerrado's non-volant mammal fauna. Biotropica 18:126-135.

Remsen, J.V., Jr. and T.A. Parker, III. 1983. Contribution of river-created habitats to bird species richness in Amazonia. Biotropica 15:223-231.

Reynel, C. and A. Gentry. In prep. Flórula de la Zona Reservada Tambopata, Perú.

Rodríguez, L.O. In press. A new species of the *Eleutherodactylus conspicillatus* group from Peru, with comments on its call. Alytes.

Rodríguez, L.O. In press. Variation in diversity of anuran faunas in Peru, upper Amazon. *In*: T. Lovejoy and G. Prance ed., Workshop 90 Proceedings.

Rodríguez, L.O. and J. Cadle. 1990. A preliminary overview of the herpetofauna of Cocha Cashu, Manu National Park, Peru. Pp. 410-425. *In*: A. Gentry (ed.) Four Neotropical Rainforests. Yale Univ. Press.

Rodríguez, L.O. and C. Myers. 1993. A new poison-frog from Manu National Park, southeastern Peru (Dendrobatidae, *Epipedobates*). Amer. Mus. Novit. 3068:1-15.

Rosenberg, G.H. 1990. Habitat specialization and foraging behavior by birds of Amazonian river islands in northeastern Peru. Condor 92:427-443.

Schaller, G.B. and M.C. Vasconcelos. 1978. A marsh deer census in Brazil. Oryx 14:341-351.

Terborgh, J., S.K. Robinson, T.A. Parker, III, C.A. Munn, and N. Pierpont. 1990. Structure and organization of an Amazonian forest bird community. Ecol. Monogr. 60:213-238.

Terborgh, J.W., J.W. Fitzpatrick and L. Emmons. 1984. Annotated checklist of bird and mammal species of Cocha Cashu Biological Station, Manu National Park, Peru. Fieldiana (Zoology, New Ser.) 21:1-29.1

TReeS (The Tambopata Reserve Society). 1993. TReeS Tambopata-Candamo Expedition 1992: A Biological Survey in the Tambopata-Candamo Reserved Zone, Madre de Dios Region, Southeast Peru. Unpublished draft, April, 1993.

Wilson, D.E. and D.M. Reeder. 1992. Mammal Species of the World. 2nd ed. Washington, DC: Smithsonian Institution Press.

Woodman, N., R.M. Timm, R. Arana C., V. Pacheco, C.A. Schmidt, E.D. Hooper, and C. Pacheco A. 1991. Annotated checklist of the mammals of Cuzco Amazonico, Peru. Occas. Pap., Mus. Nat. Hist., Univ. Kansas 145:1-12.

Wust, W. 1989. Mamíferos del albergue Tambopata Research Center y alrededores, incluyendo la boca del Río Távara. Anexo 3. *In*: Estado Actual del Conocimiento de la Zona Reservada Tambopata-Candamo. CDC-Perú and ACSS.

Gazetteer

Coordinates were taken with a Magellan GPS receiver. Several cold fronts during the expedition caused wild fluctuations of barometric pressure, making altimeter readings unreliable.

PERÚ:

Departamento de Madre de Dios, Río Tambopata

1 Explorers Inn. 12° 50.3' S; 69° 17.7' W. El. ca. 270-290 m. East bank of the Río Tambopata below the mouth of the Río de La Torre. Headquarters for studies in the Explorers Inn Reserve (the old Zona Reservada de Tambopata, or Tambopata Reserved Zone) on trail systems reaching approximately 5 km to the E.

2 Ccolpa de Guacamayos. 13° 08.5' S; 69° 36.4' W. West bank of the Río Tambopata, at Tambopata Nature Tours Lodge and Centro de Investigaciones. Immediately below ridge with the spectacular avian mineral lick. On trail systems to approx 5 km W.

Departamento de Puno, Río Távara

3 Mirador, Boca Távara. Junction of Río Tambopata and Río Távara ridge SW of junction, overlooking abandoned site of Astillero. Trail about 3 km up ridgeline to overlook facing N.

4 Cerros del Távara, Fila SE, junction Río Candamo - Río Guacamayo. 13° 30.2' S; 69° 41.0'W. El. 360 m. Base of ridge arising SE of junction of Ríos Távara, Guacamayo and Candamo, in the angle between the Ríos Távara and Guacamayo. Parker et al. worked along a trail cut approximately 4-5 km up the SE branch of the ridge from the river junction to approximately 950 m el. Gentry et al. worked along a trail cut to the same ridge top at 850 m from a basecamp on the Río Guacamayo below the first island.

5 Cerros del Távara, Fila NE, entrance to the Cañón del Tavara. The ridge branch immediately to the N of (4) and joining it to the E. Surveyed by R. Foster and W. Wust only, up ridge from river approximately 4-5 km, to 900 m.

Departamento de Madre de Dios, Río Heath

6 Puesto San Antonio. 12° 39.5' S; 68° 44.3' W. El. ca. 200 m. W bank Río Heath. New headquarters of Pampas del Heath National Sanctuary.

7 N Boundary line, Pampas del Heath National Sanctuary. A wide cut transect line E-W, approx 2 km upriver from Puesto San Antonio.

8 Refugio Picoplancha. 12° 48.4' S; 68° 49.6.9' W. El. ca. 211 m. W Bank Río Heath. A new guard station, 6 h by slow peque-peque from Puesto San Antonio.

9 Refugio Juliaca. 12° 57.4' S; 68° 52.9' W. El. ca. 200 m. W. bank Río Heath. A new guard station at head of access trail to Pampas. The trail goes straight W for about 3 km to edge of Pampas, then about 2 km across the Pampa to the tip of a tongue of gallery forest where a N flowing stream gathers from a line of *Mauritia* palm-swamp into deep clear pools.

10 "Aguas Claras" Camp. 12° 57.3' S; 68° 54.8' W. El. ca. 190 m. Gallery forest in Pampa, by first of above pools.

11 Northern end of a closed over tip of pampas accessed by 45 min walk on castañero trails from a point 1/2 h downstream from Aguas Claras. RF, WW, LHE 17 June.

12 Puerto Pardo. Small frontier town located at the mouth of the Río Heath at the Río Madre de Dios.

13 Fundo Miraflores. Located about two hours upstream from Puerto Pardo.

Appendices

Birds of the Cerros del Távara (300-900 m)

Theodore A. Parker, III and Walter Wust

	Habitats	Foraging	Sociality	Abundance	Evidence
TINAMIDAE (5)					
Tinamus tao	Fh	T	S	F	t
T. major	Ft	T	S	U	t
T. osgoodi	Fm	T	S	R	t?
Crypturellus soui	Fh	T	S	U	t
C. obsoletus	Fh	T	S	U	t
ARDEIDAE (1)					
Tigrisoma fasciatum	Rm	T,W	S	F	si
CATHARTIDAE (2)					
Cathartes aura	Fh	T	S	U	si
C. melambrotus	Fh	T	S	F	si
ACCIPITRIDAE (7)					
Elanoides forficatus	Fh,Fm	A,C	S,G	F	si
Harpagus bidentatus	Fh	C,Sc	S	U	si
Ictinia plumbea	Fh	A,C	S	U	si
Buteogallus urubitinga	Fh,Rm	T,C	S	U	si
Buteo magnirostris	Fe,Rm	T,C	S	F	si
Spizastur melanoleucus	Fh	C,A	S	R	si
Spizaetus ornatus	Fh	C,T	S	R	si
FALCONIDAE (4)					
Herpetotheres cachinnans	Fh,Rm	C,T	S	U	t
Micrastur ruficollis	Fh	U,C	S	U	t
Daptrius americanus	Fh	C	G	F	t
Falco rufigularis	Fh	A	S	R	si
CRACIDAE (4)					
Penelope jacquacu	Fh	C,T	S,G	F	t
Aburria pipile	Fh,Fm	C,T	S,G	U	si
Mitu tuberosa	Fh,Fm	T,Sc	S	F	t
Pauxi unicornis	Fm	T,Sc	S	R	si
PHASIANIDAE (1)					
Odontophorus stellatus	Fh	T	G	F	si
EURYPYGIDAE (1)					
Eurypyga helias	Rm	T	S	F	si

Habitats

Fh	Upland Forest
Ft	Floodplain forest
Fm	Montane evergreen forest
Fe	Forest edge
Fsm	Forest stream margins
B	Bamboo
R	River
Rm	River margins
O	Overhead

Foraging Position

T	Terrestrial
U	Undergrowth
Sc	Subcanopy
C	Canopy
W	Water
A	Aerial

Sociality

S	Solitary or in pairs
G	Gregarious
M	Mixed-species flocks
A	Army ant followers

Abundance

C	Common
F	Fairly common
U	Uncommon
R	Rare
(M)	Migrant, origin unknown
(Mn)	Migrant from north
(Ms)	Migrant from south

Evidence

sp	Specimen
t	Tape
si	Species ID by sight

	Habitats	Foraging	Sociality	Abundance	Evidence
PSOPHIIDAE (1)					
Psophia leucoptera	Fh	T	G	U	t
COLUMBIDAE (5)					
Columba plumbea	Fh,Fm	C	S	R	t
C. subvinacea	Fh	C	S	C	t
Leptotila rufaxilla	Fe,Ft	T	S	F	si
Geotrygon saphirina	Fm	T	S	F	t
G. montana	Fh,Fm	T	S	U	si
PSITTACIDAE (7)					
Ara macao	Fh	C	S,G	F	t
A. chloroptera	Fh,Fm	C	S	F	t
A. severa	Fe,Ft	C	S,G	U	si
Pyrrhura picta	Fh	C	G	F	t
Brotogeris cyanoptera	Fh,Ft	C	G	U	t
Pionus menstruus	Fh,Fm	C	S,G	F	t
Amazona farinosa	Fh	C	S,G	F	t
CUCULIDAE (2)					
Piaya cayana	Fh,Fm	C	S,M	F	t
Neomorphus geoffroyi	Fh	T	S,A	R	si
STRIGIDAE (5)					
Otus guatemalae	Fh,Fm	C	S	F	t
Lophostrix cristata	Fh	C	S	F	t
Pulsatrix perspicillata	Ft,Fh	C,U	S	U	t
P. melanota	Fh,Fm	C,Sc	S	U	t
Ciccaba virgata	Fh	C	S	U	t
CAPRIMULGIDAE (1)					
Hydropsalis climacocerca	Rm	A	S	U	si
APODIDAE (3)					
Streptoprocne zonaris	O(Fh,Fm)	A	G	C	si
Chaetura cinereiventris	O(Fh,Fm)	A	S,G	F	t
Cypseloides sp.	O(Fh)	A	G	U	t
TROCHILIDAE (8)					
Phaethornis superciliosus	Fh,Fm	U	S	C	sp,t
P. ruber	Fh	U	S	F	si
Eutoxeres condamini	Fm	U	S	U	sp

	Habitats	Foraging	Sociality	Abundance	Evidence
Florisuga mellivora	Fh	C,Sc	S	F	si
Klais guimeti	Fm	Sc,U	S	U	si
Thalurania furcata	Fh,Fm	U,C	S	F	si
Amazilia sp.	Fm	Sc	S	R?	si
Polyplancta aurescens	Fh	C	S	U	si
TROGONIDAE (4)					
Trogon melanurus	Fh,Ft	C,Sc	S	F	t
T. collaris	Fh	Sc	S	C	t
T. curucui	Fh	C	S	U	si
T. violaceus	Fh	C	S	F	t
MOMOTIDAE (2)					
Electron platyrhynchum	Fh	C	S	U	si
Momotus momota	Fh	Sc,C	S	U	t
ALCEDINIDAE (4)					
Ceryle torquata	Rm	W	S	U	si
Chloroceryle amazona	Rm	W	S	U	si
C. americana	Rm,Fsm	W	S	F	si
C. inda	Fsm	W	S	U	si
BUCCONIDAE (5)					
Notharchus macrorhynchos	Fh	C	S	U	si
Bucco capensis	Fh	C,Sc	S	U	t
Nystalus striolatus	Fh,Fm	C	S	F	t
Malacoptila semicincta	Fh	U,Sc	S	U	si
Monasa morphoeus	Fh	C	G	F	t
GALBULIDAE (2)					
Galbula cyanescens	Fe,Ft	Sc	S,M	U	si
Jacamerops aurea	Fh	C	S	U	si
CAPITONIDAE (2)					
Eubucco richardsoni	Fh	C	S,M	F	t
Capito niger	Fh,Fm	C	S,M	C	t
RAMPHASTIDAE (5)					
Aulacorhynchus prasinus	Fh,Fm	C,Sc	S	U	t
Pteroglossus mariae	Fh	C	S,G	F	t
Selenidera reinwardtii	Fh	C,Sc	S	U	t
Ramphastos culminatus	Fh,Fm	C	S,G	C	t

Habitats

Fh	Upland Forest
Ft	Floodplain forest
Fm	Montane evergreen forest
Fe	Forest edge
Fsm	Forest stream margins
B	Bamboo
R	River
Rm	River margins
O	Overhead

Foraging Position

T	Terrestrial
U	Undergrowth
Sc	Subcanopy
C	Canopy
W	Water
A	Aerial

Sociality

S	Solitary or in pairs
G	Gregarious
M	Mixed-species flocks
A	Army ant followers

Abundance

C	Common
F	Fairly common
U	Uncommon
R	Rare
(M)	Migrant, origin unknown
(Mn)	Migrant from north
(Ms)	Migrant from south

Evidence

sp	Specimen
t	Tape
si	Species ID by sight

	Habitats	Foraging	Sociality	Abundance	Evidence
R. cuvieri	Fh	C	S,G	F	t
PICIDAE (6)					
Picumnus aurifrons	Fh	C	M	F	t
Veniliornis affinis	Fh,Fm	C	M	F	t
Piculus leucolaemus	Fh,Fm	C	S,M	F	t
Celeus grammicus	Fh,Fm	C	S,M	U	si
Campephilus rubricollis	Fh,Fm	U,C	S	U	t
Dryocopus lineatus	Ft,Fe	C	S	U	t
DENDROCOLAPTIDAE (10)					
Deconychura longicauda	Fh,Fm	C,Sc	S,M	F	t
Sittasomus griseicapillus	Fh	C,Sc	S,M	R	si
Glyphorynchus spirurus	Fh,Fm	U,C	S,M	C	t
Xiphocolaptes promeropirhynchus	Fh	C,Sc	S,M	R	t
Dendrocolaptes picumnus	Fh,Fm	Sc	S,M	U	t
Xiphorhynchus ocellatus	Fm	U,Sc	S,M	C	sp,t
X. spixii	Fh	U,Sc	M	U	t
X. guttatus	Fh,Fm	Sc,C	S,M	C	t
Lepidocolaptes albolineatus	Fh	C	C	F	t
Campylorhamphus trochilirostris	Fh,Fm	Sc,C	S,M	U	t
FURNARIIDAE (13)					
Cranioleuca curtata	Fm	C	M	U	t
Hyloctistes subulatus	Fh,Fm	Sc,C	S,M	F	t
Ancistrops strigilatus	Fh	C	M	U	si
Philydor erythrocercus ochrogaster	Fm	C	M	U	si
P. erythropterus	Fh	C	M	F	si
P. ruficaudatus	Fh,Fm	C	M	F	t
Automolus infuscatus	Fh	U,Sc	M	U	si
A. rubiginosus	Fh,Fm	U,Sc	M	U	t
A. ochrolaemus	Fh	U,Sc	M	C	t
Xenops rutilans	Fm	C	C	F	t
X. minutus	Fh	U,Sc	M	U	si
Sclerurus albigularis	Fm	T	S	U	sp,t
S. caudacutus	Fh	T	S	U	sp,t
FORMICARIIDAE (28)					
Cymbilaimus lineatus	Fh	Sc,C	S,M	C	t

	Habitats	Foraging	Sociality	Abundance	Evidence
Thamnophilus palliatus	Fm,Fe	C,Sc	S	R	t
T. schistaceus	Fh	Sc	M,S	C	t
Pygiptila stellaris	Fh	C	M	U	si
Thamnistes anabatinus	Fm	C,Sc	M	F	t
Dysithamnus mentalis	Fm	Sc	M,S	C	t
Thamnomanes schistogynus	Fh	U,Sc	M	U	t
T. ardesiacus	Fh	U	M	R	t
Myrmotherula brachyura	Fh	C	M,S	C	t
M. sclateri	Fh	C	M	U	si
M. leucophthalma	Fh,Fm	U,Sc	M	F	t
M. erythrura	Fm	Sc,C	M	F	t
M. axillaris	Fh	U,Sc	M	C	t
M. menetriesii	Fh	Sc,C	M	C	t
Dichrozona cincta	Fh	T	S	F	sp,t
Herpsilochmus rufimarginatus	Fm	C	S,M	C	t
Terenura humeralis	Fh,Fm	C	M	U	t
Cercomacra cinerascens	Fh	C	S,M	C	t
C. serva	Fe,B	U	S	R	t
Myrmoborus leucophrys	Ft	U,T	S	U	si
M. myotherinus	Fh	U,T	S	C	t
Hypocnemis cantator	Ft,Fh	Sc,U	S,M	F	t
Myrmeciza hemimelaena	Fh	U,T	S	C	t
M. fortis	Fh	T,U	S,A	C	t
M. atrothorax	Fe	T,U	S	R	si
Rhegmatorhina melanosticta	Fh	U,T	S,A	U	t
Hylophylax naevia	Fh	U	S	C	t
H. poecilinota	Fh	U	S,A	F	t
Formicarius analis	Fh	T	S	C	t
Myrmothera campanisona	Fh	T	S	F	t
Conopophaga peruviana	Fh	U	S	F	t
TYRANNIDAE (37)					
Zimmerius gracilipes	Fh	C	S,M	C	t
Ornithion inerme	Fh	C	S,M	F	t
Tyrannulus elatus	Fh	C	S,M	R	t
Myiopagis gaimardii	Fh	C	M	C	t

Habitats

Fh	Upland Forest
Ft	Floodplain forest
Fm	Montane evergreen forest
Fe	Forest edge
Fsm	Forest stream margins
B	Bamboo
R	River
Rm	River margins
O	Overhead

Foraging Position

T	Terrestrial
U	Undergrowth
Sc	Subcanopy
C	Canopy
W	Water
A	Aerial

Sociality

S	Solitary or in pairs
G	Gregarious
M	Mixed-species flocks
A	Army ant followers

Abundance

C	Common
F	Fairly common
U	Uncommon
R	Rare
(M)	Migrant, origin unknown
(Mn)	Migrant from north
(Ms)	Migrant from south

Evidence

sp	Specimen
t	Tape
si	Species ID by sight

	Habitats	Foraging	Sociality	Abundance	Evidence
M. viridicata	Fh	C,Sc	M	U	si
Mionectes olivaceus	Fh,Fm	Sc,C	M	F	si
M. oleagineus	Fh	U,C	S,M	F	si
Phylloscartes orbitalis	Fm	C,Sc	M	F	sp,t
Phylloscartes sp. nov.	Fm	C	M	U	t
Corythopis torquata	Fh	T	S	F	t
Myiornis albiventris	Fm	C	S,M	U	t
M. ecaudatus	Fh	C	S	U	t
Hemitriccus zosterops	Fh	Sc,C	S	U	si
H. rufigularis	Fm	Sc,U	S,M	F	sp
Todirostrum chrysocrotaphum	Fh,Ft	C	S,M	C	t
Ramphotrigon ruficauda	Fh	Sc	S	U	si
Rhynchocyclus olivaceus	Fh	Sc,C	M,S	R	si
Tolmomyias assimilis	Fh	C	M	C	t
Platyrinchus platyrhynchos	Fh	Sc,U	S	F	t
Terenotriccus erythrurus	Fh	Sc,C	S,M	F	t
Myiobius (barbatus)	Fh	Sc,C	M	U	si
Myiotriccus ornatus	Fm	Sc	S,M	F	t
Pyrrhomyias cinnamomea	Fm	Sc,C	S,M	R	si
Lathrotriccus euleri	Fh,B	Sc	S	U	t
Sayornis nigricans	Rm	A	S	F	si
Ochthornis littoralis	Rm	A	S	F	si
Attila spadiceus	Fh	C,Sc	S,M	F	t
Rhytipterna simplex	Fh	C	S,M	F	t
Laniocera hypopyrra	Fh	C	S,M	U	si
Myiarchus tuberculifer	Fm	C,Sc	M	R	si
Myiozetetes luteiventris	Fh	C	S	U	t
Conopias sp.	Fm	C	M	R	si
Tyrannus melancholicus	Fe,Rm	A,C	S	F	si
Myiodynastes chrysocephalus	Fm	C	S	U	t
Pachyramphus marginatus	Fh	C	M	U	si
P. minor	Fh	C	M	U	si
Tityra semifasciata	Fh	C	M	F	si
COTINGIDAE (5)					
Pipreola chlorolepidota	Fm	C	S,M	R	si

	Habitats	Foraging	Sociality	Abundance	Evidence
Lipaugus vociferans	Fh	C,Sc	S	F	t
Cotinga cayana	Fh	C	S	F?	si
Cephalopterus ornatus	Ft,Fh	C,Sc	S	U	si
Oxyruncus cristatus	Fm	C	M	F	si
PIPRIDAE (7)					
Schiffornis turdinus	Fh,Fm	U	S	F	t
Piprites chloris	Fh,Fm	C	M	F	t
Tyranneutes stolzmanni	Fh	C,Sc	S,M	U	t
Chiroxiphia boliviana	Fm	U,C	S	R	si
Chloropipo holochlora	Fm	U,Sc	S	U	sp
Pipra coronata	Fh	U,Sc	S	C	t
P. chloromeros	Fh	U,Sc	S	C	si
HIRUNDINIDAE (3)					
Tachycineta albiventer	R	A	S	U	si
Notiochelidon cyanoleuca	R	A	S,G	F	si
Atticora fasciata	R	A	S,G	C	si
TROGLODYTIDAE (4)					
Thryothorus genibarbis	Fh,B	U	S	F	t
Microcerculus marginatus	Fh	T	S	F	t
Cyphorhinus aradus	Fh	T,U	S,M	U	t
SYLVIINAE (2)					
Ramphocaenus melanurus	Fh,Fm	Sc,C	M	F	t
Microbates cinereiventris	Fm	U	S,M	C	t
TURDINAE (1)					
Turdus albicollis	Fh,Fm	T,C	S	F	t
VIREONIDAE (3)					
Vireo olivaceus	Fh	C	M	F	si
Hylophilus hypoxanthus	Fh	C	M	C	t
H. ochraceiceps	Fh	U,Sc	M	C	t
CARDINALINAE (4)					
Caryothraustes humeralis	Fh	C	M	F	si
Pitylus grossus	Fh	C	S,M	C	t
Saltator maximus	Fh,Ft	C,Sc	S,M	U	t
Cyanocompsa cyanoides	Fh	U,Sc	S	U	sp

Habitats

Fh	Upland Forest
Ft	Floodplain forest
Fm	Montane evergreen forest
Fe	Forest edge
Fsm	Forest stream margins
B	Bamboo
R	River
Rm	River margins
O	Overhead

Foraging Position

T	Terrestrial
U	Undergrowth
Sc	Subcanopy
C	Canopy
W	Water
A	Aerial

Sociality

S	Solitary or in pairs
G	Gregarious
M	Mixed-species flocks
A	Army ant followers

Abundance

C	Common
F	Fairly common
U	Uncommon
R	Rare
(M)	Migrant, origin unknown
(Mn)	Migrant from north
(Ms)	Migrant from south

Evidence

sp	Specimen
t	Tape
si	Species ID by sight

Habitats

Fh	Upland Forest
Ft	Floodplain forest
Fm	Montane evergreen forest
Fe	Forest edge
Fsm	Forest stream margins
B	Bamboo
R	River
Rm	River margins
O	Overhead

Foraging Position

T	Terrestrial
U	Undergrowth
Sc	Subcanopy
C	Canopy
W	Water
A	Aerial

Sociality

S	Solitary or in pairs
G	Gregarious
M	Mixed-species flocks
A	Army ant followers

Abundance

C	Common
F	Fairly common
U	Uncommon
R	Rare
(M)	Migrant, origin unknown
(Mn)	Migrant from north
(Ms)	Migrant from south

Evidence

sp	Specimen
t	Tape
si	Species ID by sight

	Habitats	Foraging	Sociality	Abundance	Evidence
THRAUPINAE (22)					
Lamprospiza melanoleuca	Fh	C	G,M	F	si
Chlorospingus canigularis	Fm	C	M	F	si
Hemithraupis sp.	Fh	C	M	U	si
Chlorothraupis carmioli	Fm	U,C	G,M	C	sp,t
Lanio versicolor	Fh,Fm	C	M	C	t
Tachyphonus rufiventer	Fh,Fm	C	M	F	t
T. luctuosus	Fh	C,Sc	M	U	si
Piranga flava	Fm	C	M	U	si
Ramphocelus carbo	Fe,Ft	U,C	G,M	F	si
Thraupis palmarum	Fh	C	S,M	U	si
Euphonia xanthogaster	Fh,Fm	C,U	M,S	F	sp,t
E. rufiventris	Fh	C	M,S	U	t
Chlorophonia cyanea	Fm	C	M,S	F	t
Tangara chilensis	Fh,Fm	C	M,G	C	t
T. schrankii	Fh	C	M	F	si
T. arthus	Fm	C	M	U	t
T. xanthogastra	Fh,Fm	C	M	C	si
T. gyrola	Fh,Fm	C	M	C	si
T. cyanicollis	Fh,Fm	C	M	F	t
Dacnis lineata	Fh,Fm	C	M	F	t
D. cayana	Fh,Fm	C	M	C	si
Cyanerpes caeruleus	Fh	C	M	F	sp
PARULIDAE (4)					
Parula pitiayumi	Fm	C	M	C	t
Myioborus miniatus	Fm	C	M	C	t
Basileuterus chrysogaster	Fm	C	M	F	t
Phaeothlypis rivularis	Ft,Fsm	T,U	S	U	t
ICTERIDAE (4)					
Psarocolius angustifrons	Fh,Fm	C	G,M	F	t
P. yuracares	Fh	C	G,S	F	si
Clypicterus oseryi	Fh	C	S,G	U	t
Cacicus cela	Fh	C,Sc	G,M	F	t
Total:	**234 species**				
Montane Forest Component:	ca. 37 species				

Birds of the Ccolpa de Guacamayos, Madre de Dios

APPENDIX 2

Theodore A. Parker, III, Andrew W. Kratter and Walter Wust

	Habitats	Foraging	Sociality	Abundance	Evidence
TINAMIDAE (7)					
Tinamus major	Ft	T	S	F	t
T. guttatus	Fh,Ft,B	T	S	F	t
Crypturellus cinereus	Ft	T	S	C	t
C. soui	Ft	T	S	F	t
C. undulatus	Z,Ft,B	T	S	C	t
C. bartletti	Ft	T	S	U	t
C. atrocapillus	Ft,B	T	S	F	t
PHALACROCORACIDAE (1)					
Phalacrocorax brasilienisis	R	W	S,G	U	si
ANHINGIDAE (1)					
Anhinga anhinga	L,R	W	S	U	ph
ARDEIDAE (8)					
Pilherodias pileatus	S,Rm	W	S	F	ph,t
Ardea cocoi	S,Rm	W	S	U	ph
Casmerodius albus	S	W	S	R	si
Bubulcus ibis	S	T	S,G	R	si
Egretta thula	S,Rm	W	S	F	si
E. caerulea	Rm	W	S	R	si
Butorides striatus	M	W	S	R	si
Tigrisoma lineatum	Fsm,Fs	W	S	U	ph,t
CICONIIDAE (2)					
Mycteria americana	S	W	G,S	R	si
Jabiru mycteria	S	W	S	U	si
THRESKIORNITHIDAE (1)					
Ajaia ajaja	S	W	S	R	si
ANHIMIDAE (1)					
Anhima cornuta	S	W	S	R	si
ANATIDAE (2)					
Neochen jubata	S	T	S	R	ph
Cairina moschata	R,Fs	W	S	U	si
CATHARTIDAE (4)					
Sarcoramphus papa	Ft	T	S	U	ph

Habitats

Fh	Upland Forest
Ft	Floodplain forest
Fs	Swamp forest
Fe	Forest edge
Fsm	Forest stream margins
Fo	Forest openings
Z	"Zabolo"
B	Bamboo
R	River
Rm	River margins
S	Shores
M	Marshes
A	Aguajales
O	Overhead

Foraging Position

T	Terrestrial
U	Undergrowth
Sc	Subcanopy
C	Canopy
W	Water
A	Aerial

Sociality

S	Solitary or in pairs
G	Gregarious
M	Mixed-species flocks
A	Army ant followers

Abundance

C	Common
F	Fairly common
U	Uncommon
R	Rare
(M)	Migrant
(Mn)	Migrant from north
(Ms)	Migrant from south
(Mm)	Migrant from Andes

Evidence

sp	Specimen
t	Tape
si	Species ID by sight
ph	Photo

	Habitats	Foraging	Sociality	Abundance	Evidence
Coragyps atratus	S,Rm,Ft,Fe	T	S,G	U	ph
Cathartes aura	S,Rm,Ft	T	S	R	ph
Cathartes melambrotus	Z,Ft,Fh	T	S	F	si
ACCIPITRIDAE (17)					
Elanoides forficatus	Ft	A	S,G	U	si
Leptodon cayanensis	Ft,Fh	C,Sc	S	R	ph,t
Harpagus bidentatus	Fh	Sc	S	U	t?
Ictinia plumbea	Z,Ft	A	S,G	U	si
Accipiter bicolor	Ft	U,Sc	S	R	si
A. superciliosus	Ft,Fh	C,Sc	S	R	ph,t
Buteo magnirostris	Z,Rm,Fe	T,C	S	F	ph,t
B. brachyurus	Ft	A,C	S	U	si
Leucopternis kuhli	Ft,Fh	C	S	R	si
L. schistacea	Fs,Fsm,Ft	T,Sc	S	U	si
Buteogallus urubitinga	Rm,Lm,S	T,U	S	U	ph,t
Busarellus nigricollis	Fs,Fsm,Lm	T,U	S	R	si
Morphnus guianensis	Ft,Fh	Sc,C	S	R	si
Spizastur melanoleucus	Ft	C,A?	S	R	si
Spizaetus ornatus	Ft,Fh	C,T	S	R	t?
Spizaetus tyrannus	Z,Ft	Sc,C	S	U	t?
Geranospiza caerulescens	Ft,Fs	T,C	S	R	si
PANDIONIDAE (1)					
Pandion haliaetus	R	W	S	R(Mn)	si
FALCONIDAE (8)					
Herpetotheres cachinnans	Ft,Fh	C,T	S	F	t
Micrastur semitorquatus	Ft	Sc,U	S	R	si
M. mirandollei	Ft,Fh	Sc,U	S	R	si
M. ruficollis	Ft,Fh	Sc,U	S	U	t
M. gilvicollis	Ft,Fh	Sc,U	S	U	t
Daptrius ater	Rm,S,Z	T,C	S,G	F	t
D. americanus	Ft,Fh	C	G	F	t
Falco rufigularis	Rm,Fe	A	S	F	ph,t
CRACIDAE (4)					
Ortalis guttata	Z,Ft	C	G	U	ph,t
Penelope jacquacu	Ft,Fh	C,T	S,G	F	ph,t

	Habitats	Foraging	Sociality	Abundance	Evidence
Aburria pipile	Ft,Z	C	S	U	ph,t
Mitu tuberosa	Ft,Fh,Fsm	T	S	R	t
PHASIANIDAE (1)					
Odontophorus stellatus	Ft,Fh	T	G	U	t
OPISTHOCOMIDAE (1)					
Opisthocomus hoazin	Lm,Rm	U,C	G	U	ph,t
PSOPHIIDAE (1)					
Psophia leucoptera	Fh,Ft	T	G	U	t
RALLIDAE (1)					
Aramides cajanea	Fsm,Fs	T	S	F	t
EURYPYGIDAE (1)					
Eurypyga helias	Fsm,Fs,S	T,W	S	U	si
CHARADRIIDAE (2)					
Hoploxypterus cayanus	S	T	S	R	si
Charadrius collaris	S	T	S	U	si
SCOLOPACIDAE (5)					
Tringa solitaria	S	T	S	U(Mn)	ph
T. flavipes	S	T	S	R(Mn)	si
Actitis macularia	S	T	S	U(Mn)	t
Calidris fuscicollis	S	T	S	R(Mn)	si
C. himantopus	S	T	S	R(Mn)	si
LARIDAE (2)					
Phaetusa simplex	R	W,A	S	U	t
Sterna superciliaris	R	W,A	S	F	t
RHYNCHOPIDAE (1)					
Rynchops niger	R	W	S,G	R	si
COLUMBIDAE (7)					
Columba speciosa	Rm	C	S	R	si
C. cayennensis	Z	U,C	S,G	C	ph,t
C. subvinacea	Ft	C	S	F	t
C. plumbea	Ft,Fh	C	S	C	t
Columbina talpacoti	Rm,Fe	T	S,G	R	si
Leptotila rufaxilla	Z,Fe	T	S	F	t
Geotrygon montana	Ft,Fh	T	S	F	t

Habitats

Fh	Upland Forest
Ft	Floodplain forest
Fs	Swamp forest
Fe	Forest edge
Fsm	Forest stream margins
Fo	Forest openings
Z	"Zabolo"
B	Bamboo
R	River
Rm	River margins
S	Shores
M	Marshes
A	Aguajales
O	Overhead

Foraging Position

T	Terrestrial
U	Undergrowth
Sc	Subcanopy
C	Canopy
W	Water
A	Aerial

Sociality

S	Solitary or in pairs
G	Gregarious
M	Mixed-species flocks
A	Army ant followers

Abundance

C	Common
F	Fairly common
U	Uncommon
R	Rare
(M)	Migrant
(Mn)	Migrant from north
(Ms)	Migrant from south
(Mm)	Migrant from Andes

Evidence

sp	Specimen
t	Tape
si	Species ID by sight
ph	Photo

	Habitats	Foraging	Sociality	Abundance	Evidence
PSITTACIDAE (18)					
Ara ararauna	A,Fs,Ft	C	G	C	ph,t
A. macao	Ft,Fh	C	S,G	C	ph,t
A. chloroptera	Ft,Fh	C	S,G	C	ph,t
A. severa	Z,Ft	C	S,G	C	ph,t
A. manilata	Ft	C	G	C	ph,t
A. couloni	Ft	C	S,G	F	t
Aratinga leucophthalmus	Z,Ft	C	G	C	ph,t
A. weddellii	Z,Ft,R	C	G	C	ph,t
Pyrrhura rupicola	Ft,Fh	C	G	C	t
P. picta	Ft	C	G	U	si
Forpus sclateri	Ft,Fh	C	S	F	t
Brotogeris cyanoptera	Z,Ft,Fh	C	G	C	t
Touit huetii	?(O)	C	G	R	t
Nannopsittaca dachilleae	Ft	C	S,G	F	t
Pionites leucogaster	Ft,Fh	C	G	F	t
Pionopsitta barrabandi	Fh,Ft	Sc,C	S,G	C	t
Pionus menstruus	Ft	C	S,G	C	t
Amazona ochrocephala	Ft,Z	C	S	C	t
A. farinosa	Ft,Fh	C	S,G	C	t
CUCULIDAE (6)					
Coccyzus cinereus	Ft,W,Z	C	M,S	R(Ms)	si
Piaya cayana	Z,Ft,Fh	Sc,C	S,M	F	ph,t
P. melanogaster	Ft,Fh	C	S,M	U	t?
Crotophaga ani	C,Lm	U,T	G	F	t?
Dromococcyx pavoninus	Ft,B	T,U	S	R	t
D. phasianellus	Ft	T,U	S	R	t
STRIGIDAE (7)					
Otus choliba	Z	Sc,C	S	R	t
O. watsonii	Ft,Fh	U,C	S	C	t
Lophostrix cristata	Ft,Fh	C	S	U	t
Pulsatrix perspicillata	Ft	C,T	S	U	t
Glaucidium hardyi	Ft,Fh	C	S	F	t
G. brasilianum	Z	C,Sc	S	U	t?
Ciccaba virgata	Ft,Fh	C	S	U	si

	Habitats	Foraging	Sociality	Abundance	Evidence
NYCTIBIIDAE (3)					
Nyctibius grandis	Ft	C,A	S	U	t
N. aethereus	Ft	C,A	S	R	t
N. griseus	Z,Ft,Fe	C,A	S	U	t?
CAPRIMULGIDAE (6)					
Lurocalis semitorquatus	O(Ft)	A	S	R	t?
Chordeiles rupestris	S	A	G	C	ph
Nyctidromus albicollis	Z,Fe	A	S	F	t
Nyctiphrynus ocellatus	Ft,B	A,Sc	S	U	t
Caprimulgus sp.	Ft	A	S	R	ph
Hydropsalis climacocerca	Rm,S	A	S	F	t?
APODIDAE (6)					
Streptoprocne zonaris	O(Ft,Fh)	A	G	F	si
Chaetura cinereiventris	O(Ft)	A	G,M	C(M?)	si
C. egregia	O(Ft)	A	G,M	U(Ms?)	si
C. brachyura	O(Fe)	A	G,M	U	si
Panyptila cayennensis	O(Ft)	A	S,M	U	si
Reinarda squamata	O(A,L,R)	A	S,G	C	si
TROCHILIDAE (15)					
Glaucis hirsuta	Z,Ft	U,Sc	S	U	si
Threnetes leucurus	Z,Ft	U	S	F	si
Phaethornis philippi	Fh	U	S	F	si
P. hispidus	Z,Ft	U	S	C	sp
P. ruber	Ft,Fh,B	U	S	F	t?
Campylopterus largipennis	Ft,Z	U,Sc	S	R	si
Florisuga mellivora	Ft,Fh	C	S	U	ph
Anthracothorax nigricollis	Rm	C	S	R	si
Lophornis sp.	Fe?	C	S	R	si
Chlorostilbon mellisugus	Fe?	U	S	R	si
Thalurania furcata	Ft,Fh	U,C	S	F	sp,t
Hylocharis cyanus	Ft	C	S	F	ph,t
Chrysuronia oenone	Fe,W	C	S	R	ph
Polyplancta aurescens	Ft	Sc,C	S	U	ph,t
Heliomaster longirostris	Z,Rm	C	S	R	si

Habitats

Fh	Upland Forest
Ft	Floodplain forest
Fs	Swamp forest
Fe	Forest edge
Fsm	Forest stream margins
Fo	Forest openings
Z	"Zabolo"
B	Bamboo
R	River
Rm	River margins
S	Shores
M	Marshes
A	Aguajales
O	Overhead

Foraging Position

T	Terrestrial
U	Undergrowth
Sc	Subcanopy
C	Canopy
W	Water
A	Aerial

Sociality

S	Solitary or in pairs
G	Gregarious
M	Mixed-species flocks
A	Army ant followers

Abundance

C	Common
F	Fairly common
U	Uncommon
R	Rare
(M)	Migrant
(Mn)	Migrant from north
(Ms)	Migrant from south
(Mm)	Migrant from Andes

Evidence

sp	Specimen
t	Tape
si	Species ID by sight
ph	Photo

	Habitats	Foraging	Sociality	Abundance	Evidence
TROGONIDAE (6)					
Pharomachrus pavoninus	Fh,Ft	C	S	U	t
Trogon melanurus	Ft,Fh	C	S	C	t
T. viridis	Ft,Fh	C,Sc	S,M	C	t
T. collaris	Ft,Fh	Sc,U	S,M	F	ph,t
T. curucui	Ft,Z	C	S,M	U	ph,t
T. violaceus	Ft,Fh	C	S,M	U	t?
ALCEDINIDAE (5)					
Ceryle torquata	Rm,Lm	W	S	F	t?
Chloroceryle amazona	Rm,Lm	W	S	F	ph,t
C. americana	Fsm,Rm,Lm	W	S	U	si
C. inda	Fsm,Lm	W	S	U	si
C. aenea	Fsm,Lm	W	S	U	si
MOMOTIDAE (3)					
Electron platyrhynchum	Ft,Fh	C,Sc	S	C	t?
Barypthengus martii	Ft	Sc,C	S	F	t?
Momotus momota	Ft,Fh	Sc,C	S	U	ph,t
GALBULIDAE (3)					
Brachygalba albogularis	Z	C,A	S	F	ph,t
Galbula cyanescens	Ft,Fo	U,Sc	S,M	F	ph,t
Jacamerops aurea	Ft,Fh	C,Sc	S	U	t?
BUCCONIDAE (10)					
Notharchus macrorhynchos	Ft,Fh	C	S	R	si
Bucco macrodactylus	Z	Sc,C	S	R	sp
B. capensis	Ft	Sc,C	S	U	si
Nystalus striolatus	Ft,Fh	C,	S	U	t
Malacoptila semicincta	Ft,Fh	U	S	U	t?
Nonnula ruficapilla	Ft,Z,B	U,Sc	S	U	t
Monasa nigrifrons	Ft,Z,	C,Sc	G,M	C	ph,t
M. morphoeus	Ft,Fh	C,Sc	G,M	C	t
M. flavirostris	Ft	C	S,M	U	t
Chelidoptera tenebrosa	Rm,Z,Ft	C,A	S	F	t?
CAPITONIDAE (3)					
Capito niger	Ft,Fh	C	S,M	C	t
Eubucco richardsoni	Ft	C	S,M	F	ph,t

	Habitats	Foraging	Sociality	Abundance	Evidence
E. tucinkae	Z,Ft	C	S,M	R	t
RAMPHASTIDAE (8)					
Aulacorhynchus prasinus	Ft	C	S	U	sp,t
Pteroglossus castanotis	Ft,Z	C	G	F	ph,t
P. inscriptus	Ft	C	G	R	t
P. mariae	Ft,Fh	C	G	F	t
P. beauharnaesii	Ft,Fh	C	G	U	t
Selenidera reinwardtii	Ft, Fh	C,Sc	S	F	t
Ramphastos culminatus	Ft,Fh	C	S,G	C	t
R. cuvieri	Ft,Fh	C	S,G	C	ph,t
PICIDAE (13)					
Picumnus rufiventris	Z,B	U	S,M	U	sp
P. aurifrons	Ft,Fh	C	S,M	U	si
Piculus leucolaemus	Ft	C	M	U	t?
Celeus elegans	Ft,Fs	C,Sc	S,M	R	sp,t
C. spectabilis	Z,Ft,B	U,Sc	S,M	U	sp,t
C. flavus	Ft	C,Sc	S,G	U	t
C. torquatus	Ft,Fh	C,Sc	S	R	t?
Dryocopus lineatus	Z,Ft	Sc,C	S	F	t
Melanerpes cruentatus	Ft,Fh,Fo	C	S,M	C	t
Veniliornis passerinus	Z	Sc,C	S,M	F	t?
V. affinis	Ft,Fh	C	M	F	t
Campephilus melanoleucus	Ft,Z	C,Sc	S	F	t
C. rubricollis	Ft,Fh	U,Sc	S	F	t
DENDROCOLAPTIDAE (15)					
Dendrocincla fuliginosa	Ft	U,Sc	S,M,A	F	t
D. merula	Ft,Fh	U,Sc	S,A	R	si
Deconychura longicauda	Ft,Fh	Sc	S,M	U	t
Sittasomus griseicapillus	Ft	U,C	M	F	t
Glyphorynchus spirurus	Ft,Fh	U,Sc	S,M	C	sp
Nasica longirostris	Lm,Fs,Ft	Sc,C	S	U	t
Dendrexetastes rufigula	Ft,Fh	C,Sc	S,M	F	t
Xiphocolaptes promeropirhynchus	Fh	U,C	S,M	R	t
Dendrocolaptes certhia	Ft,Fh	Sc,C	S,M,A	F	t
D. picumnus	Ft,Fh	U,Sc	S,A	U	t?

Habitats

Fh	Upland Forest
Ft	Floodplain forest
Fs	Swamp forest
Fe	Forest edge
Fsm	Forest stream margins
Fo	Forest openings
Z	"Zabolo"
B	Bamboo
R	River
Rm	River margins
S	Shores
M	Marshes
A	Aguajales
O	Overhead

Foraging Position

T	Terrestrial
U	Undergrowth
Sc	Subcanopy
C	Canopy
W	Water
A	Aerial

Sociality

S	Solitary or in pairs
G	Gregarious
M	Mixed-species flocks
A	Army ant followers

Abundance

C	Common
F	Fairly common
U	Uncommon
R	Rare
(M)	Migrant
(Mn)	Migrant from north
(Ms)	Migrant from south
(Mm)	Migrant from Andes

Evidence

sp	Specimen
t	Tape
si	Species ID by sight
ph	Photo

	Habitats	Foraging	Sociality	Abundance	Evidence
Xiphorhynchus obsoletus	A	U,Sc	S,M	R	t
X. spixii	Ft,Fh	U,Sc	M	C	sp
X. guttatus	Ft,Fh,Fe	Sc,C	S,M	C	ph,t
Lepidocolaptes albolineatus	Ft,Fh	C	M	F	t?
Campylorhamphus trochilirostris	Ft,B	U	S,M	U	sp
FURNARIIDAE (25)					
Furnarius leucopus	Z,Rm	T	S	F	t?
Synallaxis cabanisi	B	U	S	R	sp
S. albigularis	Rm,Fe	U	S	F	t
S. gujanensis	Z	U	S	F	t
S. rutilans	Ft,Fh	T,U	S	F	t
Cranioleuca gutturata	Ft	C,Sc	M	U	t?
Thripophaga fusciceps	Ft	U,Sc	S,M	R	t?
Berlepschia rikeri	A	C	S	R	t?
Hyloctistes subulatus	Ft	Sc,C	S,M	U	t
Ancistrops strigilatus	Ft,Fh	Sc,C	M	C	t
Simoxenops ucayale	Ft,B,Z	U	M,S	U	sp
Philydor erythrocercus	Fh,Ft	Sc	M	F	t?
P. pyrrhodes	Ft,Fh	Sc,U	S,M	U	sp,t
P. rufus	Z	C	S,M	U	t?
P. erythropterus	Ft,Fh	C	M	F	t?
P. ruficaudatus	Ft	Sc,C	M	U	t?
Automolus infuscatus	Ft,Fh	U	M	C	t?
A. dorsalis	Ft,Z,B	Sc	S,M	F	sp,t
A. rubiginosus	Ft,B	U	S	U	sp,t
A. ochrolaemus	Ft,Fh	U	S,M	F	sp,t
A. rufipileatus	Z,Ft	U	S,M	F	t
A. melanopezus	Ft,B	U,Sc	S,M	F	t
Xenops milleri	Ft,Fh	C	M	F	t?
X. minutus	Ft,Fh	U,Sc	M	F	t?
Sclerurus caudacutus	Ft,Fh	T	S	F	t
FORMICARIIDAE (47)					
Cymbilaimus lineatus	Ft,Fh	Sc,C	S,M	F	t
C. sanctaemariae	Ft,B	Sc,C	S,M	C	t
Taraba major	Z,Lm	U	S	F	t

	Habitats	Foraging	Sociality	Abundance	Evidence
Thamnophilus doliatus	Rm,Fe	U	S,M	U	t?
T. aethiops	Ft,Fh	U	S	F	t?
T. schistaceus	Ft,Fh	Sc	M	C	t
Pygiptila stellaris	Ft,Fh	C,Sc	M	C	t?
Thamnomanes ardesiacus	Ft,Fh	U	M	F	t
T. schistogynus	Ft,B	U,Sc	M	C	sp,t
Myrmotherula brachyura	Ft,Fh	C	M	C	t
M. sclateri	Ft,Fh	C	M	F	t
M. hauxwelli	Ft,Fh	U	S,M	F	t
M. leucophthalma	Ft,Fh	U	M	F	t
M. ornata	Ft,B	U,Sc	M	F	sp,t
M. axillaris	Ft,Fh,B	U,Sc	M	C	t
M. iheringi	Ft,B	U	M	U	sp,t
M. longipennis	Fh,Ft	U,Sc	M	F	t
M. menetriesii	Ft,Fh	Sc,C	M	C	t
Dichrozona cincta	Fh	T	S	R	t
Microrhopias quixensis	Ft,B	Sc	M	C	sp,t
Drymophila devillei	Ft,B	Sc	S,M	F	t
Terenura humeralis	Ft	C	M	U	t
Cercomacra cinerascens	Ft,Fh	C,Sc	S,M	C	t
C. nigrescens	Z,B	U	S	U	sp,t
C. serva	Fh,Ft,Fo	U	S	R	t
C. manu	Z,B	U	S	C	t
Myrmoborus leucophrys	Ft	U	S	F	t
M. myotherinus	Ft,Fh	U	S,A	C	t
Hypocnemis cantator	Ft,B	U	S,M	C	sp,t
Hypocnemoides maculicauda	Fs,Lm,Fsm	U,T	S	F	t
Percnostola lophotes	Z,B	T,U	S	F	sp,t
Sclateria naevia	Fs,Fsm,Lm	T	S	U	t
Myrmeciza hemimelaena	Ft,Fh,B	U,T	S	C	t
M. hyperythra	Fs,Lm	T,U	S	F	t
M. goeldii	Ft,B	U	S,A	C	sp,t
M. fortis	Ft	T	S,A	R	t?
M. atrothorax	Z,Fe	U	S	U	t
Gymnopithys salvini	Ft,Fh	U	A	F	ph

Habitats

Fh	Upland Forest
Ft	Floodplain forest
Fs	Swamp forest
Fe	Forest edge
Fsm	Forest stream margins
Fo	Forest openings
Z	"Zabolo"
B	Bamboo
R	River
Rm	River margins
S	Shores
M	Marshes
A	Aguajales
O	Overhead

Foraging Position

T	Terrestrial
U	Undergrowth
Sc	Subcanopy
C	Canopy
W	Water
A	Aerial

Sociality

S	Solitary or in pairs
G	Gregarious
M	Mixed-species flocks
A	Army ant followers

Abundance

C	Common
F	Fairly common
U	Uncommon
R	Rare
(M)	Migrant
(Mn)	Migrant from north
(Ms)	Migrant from south
(Mm)	Migrant from Andes

Evidence

sp	Specimen
t	Tape
si	Species ID by sight
ph	Photo

	Habitats	Foraging	Sociality	Abundance	Evidence
Hylophylax naevia	Ft	U	S	F	t
H. poecilinota	Ft,Fh	U	S,A	U	t?
Phlegopsis nigromaculata	Ft	T,U	S,A	U	t
Chamaeza nobilis	Ft	T	S	R	t?
Formicarius colma	Fh	T	S	F	ph,t
F. analis	Ft,Fh	T	S,A	C	t
F. rufifrons	Fe,B,Ft	T	S	U	sp
Hylopezus berlepschi	Z,B,Fo	T	S	R	t
Myrmothera campanisona	Ft,Fh	T	S	F	t
Conopophaga peruviana	Ft	U,T	S	R	t
COTINGIDAE (5)					
Lipaugus vociferans	Ft,Fh	Sc,C	S	C	t
Cotinga maynana	Ft,Z	C	S	F	si
C. cayana	Ft	C	S	R	si
Gymnoderus foetidus	Ft	C	S	F	si
Querula purpurata	Fh,Ft	C	G	C	t
PIPRIDAE (5)					
Piprites chloris	Ft,Fh	C	M	F	t
Tyranneutes stolzmanni	Ft,Fh	Sc	S	C	t
Machaeropterus pyrocephalus	Ft,B	Sc,C	S	F	ph,t
Pipra fasciicauda	Ft	U,Sc	S	F	sp,t
P. chloromeros	Ft,Fh	U,Sc	S	F	t
TYRANNIDAE (72)					
Zimmerius gracilipes	Ft,Fh	C	S,M	F	t?
Ornithion inerme	Ft,Fh	C	S,M	F	t
Camptostoma obsoletum	Z	C	S,M	U	t?
Sublegatus modestus	W,Fe	Sc,C	S,M	R(M?)	si
Phaeomyias murina	Fe	C	S,M	U	t?
Tyrannulus elatus	Ft,Fh	C,Sc	S,M	F	t?
Myiopagis gaimardii	Ft,Fh	C	M	C	t
M. caniceps	Ft	C	M	R	t?
M. viridicata	Ft,Fh,Z	Sc,C	M	R(Ms)	si
Elaenia spectabilis	Z	C,Sc	S,M	U(Ms)	si
E. parvirostris	Z	C	S,M	R(Ms)	si
E. gigas	Ft,Z	C,Cs	S,M	R	ph

	Habitats	Foraging	Sociality	Abundance	Evidence
Inezia inornata	Z	C	G,M	F(Ms)	t?
Mionectes olivaceus	Ft,Fh	C,Sc	M	R	sp
M. oleagineus	Ft,Fh	U,C	M	F	t
M. macconnelli	Ft,Fh	U,Sc	M	R	ph
Leptopogon amaurocephalus	Ft,B	U,Sc	M,S	F	sp,t
Corythopis torquata	Ft,Fh	T	S	F	t
Myiornis ecaudatus	Ft	C,Sc	S	F	t?
Poecilotriccus albifacies	B,Ft	U,Sc	M,S	U	t
Hemitriccus flammulatus	Ft,B	Sc	S	F	sp,t
H. zosterops	Ft,Fh	Sc	S	U	t
H. iohannis	Z	Sc,C	S	U	t
Todirostrum latirostre	Z,B,Rm,Fe	U	S	F	t
T. maculatum	Z	C,Sc	S,M	F	t?
T. chrysocrotaphum	Ft,Fh	C	S,M	F	t?
Ramphotrigon megacephala	Z,Ft,B	Sc,U	S,M	F	sp,t
R. fuscicauda	Ft,B	Sc	S	U	sp,t
R. ruficauda	Ft,Fh	Sc	S,M	F	t
Rhynchocyclus olivaceus	Ft	U,Sc	M	U	t
Tolmomyias assimilis	Ft,Fh	C	M	F	t?
T. poliocephalus	Ft,Z	C	S,M	F	t?
T. flaviventris	Z	C,Sc	S,M	F	t
Platyrinchus coronatus	Ft	Sc,U	S	F	t
P. platyrhynchos	Fh	U,Sc	S	F	t
Onychorhynchus coronatus	Ft,Fh,Fsm	U,Sc	S,M	U	t?
Terenotriccus erythrurus	Ft,Fh	U,Sc	S,M	U	sp,t
Myiophobus fasciatus	Fe,Z	U	S,M	U	sp,t
Contopus virens	Fe,Z	C,A	S	R(Mn)	si
C. cinereus	Z	C,A	S	R(M?)	t
Lathrotriccus euleri	Ft,Z,B	U,Sc	S	F	sp,t
Pyrocephalus rubinus	Rm,Fe	C,A	S	F(Ms)	si
Ochthoeca littoralis	Rm,S	T,A	S	F	t?
Muscisaxicola fluviatilis	Rm,S	T	S	U	si
Colonia colonus	Rm,Z	C	S	R(Mm?)	ph,t
Satrapa icterophrys	Rm	Sc	S,M	R(Ms)	si
Attila bolivianus	Ft	Sc,C	S	U	t

Habitats

Fh	Upland Forest
Ft	Floodplain forest
Fs	Swamp forest
Fe	Forest edge
Fsm	Forest stream margins
Fo	Forest openings
Z	"Zabolo"
B	Bamboo
R	River
Rm	River margins
S	Shores
M	Marshes
A	Aguajales
O	Overhead

Foraging Position

T	Terrestrial
U	Undergrowth
Sc	Subcanopy
C	Canopy
W	Water
A	Aerial

Sociality

S	Solitary or in pairs
G	Gregarious
M	Mixed-species flocks
A	Army ant followers

Abundance

C	Common
F	Fairly common
U	Uncommon
R	Rare
(M)	Migrant
(Mn)	Migrant from north
(Ms)	Migrant from south
(Mm)	Migrant from Andes

Evidence

sp	Specimen
t	Tape
si	Species ID by sight
ph	Photo

	Habitats	Foraging	Sociality	Abundance	Evidence
A. spadiceus	Ft,Fh	Sc,C	S,M	F	t
Rhytipterna simplex	Ft,Fh	Sc,C	S,M	C	t
Laniocera hypopyrra	Ft,Fh	Sc,C	S,M	F	t
Sirystes sibilator	Ft,Fh	C,Sc	S,M	F	t?
Myiarchus swainsoni	Z,Ft	C	M	U(Ms)	t?
M. ferox	Z,Fe	Sc,C	S,M	F	t?
M. tyrannulus	Ft,Fh	C	S,M	F(Ms)	t?
Pitangus sulphuratus	Rm,Fe	U,Sc	S	F	t?
Megarhynchus pitangua	Fe,Z	Sc,C	S	F	t?
Myiozetetes cayanensis	M	U	S	R	t?
M. similis	Fe,Rm	U,Sc,C	S,G	C	t?
M. granadensis	Z,Fe	Sc,C	S,G	F	t
M. luteiventris	Ft	C	S,G	U	t?
Myiodynastes maculatus	Ft	Sc,C	M	U(Ms)	si
Legatus leucophaius	Ft,Fo,Lm	C	S	F	t?
Empidonomus varius	Ft	C	M	R(Ms)	si
E. aurantioatrocristatus	Ft,Fh,Fo	C	S,M	F(Ms)	ph
Tyrannus melancholicus	Rm,Fe	C,A	S	F	t?
T. savanna	Rm,Lm	C	G	U(Ms?)	si
Pachyramphus polychopterus	Z,Fe	C	S,M	F	ph,t
P. marginatus	Ft,Fh	C	M	F	t?
P. minor	Ft,Fh	C	M	U	t?
Tityra cayana	Ft	C	S	F	si
T. semifasciata	Ft,Fh,Fe	C	S	U	t?
T. inquisitor	Ft	C	S	R	si
HIRUNDINIDAE (6)					
Tachycineta albiventer	R	A	S,G	C	t?
Progne tapera	R	A	G,S	U	t?
Notiochelidon cyanoleuca	R	A	G	R(Ms?)	si
Atticora fasciata	R,Fe	A	G	C	t?
Stelgidopteryx ruficollis	R,Fe	A	G	C	t?
Petrochelidon pyrrhonota	R	A	G	R(Mn)	si
CORVIDAE (2)					
Cyanocorax violaceus	Ft	Sc,C	G,M	C	t
C. cyanomelas	Z,Ft	Sc,C	G,M	U	t

	Habitats	Foraging	Sociality	Abundance	Evidence
TROGLODYTIDAE (6)					
Campylorhynchus turdinus	Fe,Z	Sc,C	S,M	F	t?
Thryothorus genibarbis	Z,Ft	U	S	C	sp,t
T. leucotis	Fe	U	S	U	si
Troglodytes aedon	Fe	U	S	F	t?
Microcerculus marginatus	Ft,Fh	T	S	F	t
Cyphorhinus arada	Ft,Fh	T	S,M	F	t
TURDIDAE (6)					
Catharus ustulatus	Ft,Z	T,U	S	R(Mn)	si
Turdus amaurochalinus	Z,Fe,Ft	T,Sc,C	S	F(Ms?)	t?
T. ignobilis	Z,W,Fe	T,Sc,C	S,M	U	si
T. lawrencii	Ft	T,Sc,C	S	F	t
T. hauxwelli	Ft	T,Sc	S	U	t
T. albicollis	Ft,Fh	T,Sc	S	F	t
VIREONIDAE (5)					
Cyclarhis gujanensis	Z	C	S,M	R	t?
Vireo olivaceus	Ft,Fh,Z	C	M	C(Ms)	t?
Hylophilus thoracicus	Ft	C	M	R?	si
H. hypoxanthus	Ft,Fh	C	M	C	t
H. ochraceiceps	Fh	U,Sc	M	R	t
EMBERIZINAE (7)					
Myospiza aurifrons	S,Fe	T,U	S	C	t?
Sporophila schistacea	Ft,B	U,Sc	S	R	sp
S. luctuosa	W	U	G,M	R(Mm?)	si
S. caerulescens	M,Fe	T,U	G,M	C(Ms)	si
S. castaneiventris	Fe	U,C	S,M	U	si
Arremon taciturnus	Ft,Fh	T,U	S	F	t?
Paroaria gularis	Rm	U,Sc,C	S,M	F	si
CARDINALINAE (5)					
Caryothraustes humeralis	Ft,Fh	C	M	R	si
Pitylus grossus	Ft	Sc	S,M	C	t
Saltator maximus	Z,Ft,Fh	Sc,C	S,M	F	t
S. coerulescens	Z,Fe	U,Sc,C	S	F	t?
Cyanocompsa cyanoides	Ft	U	S	F	t

Habitats

Fh	Upland Forest
Ft	Floodplain forest
Fs	Swamp forest
Fe	Forest edge
Fsm	Forest stream margins
Fo	Forest openings
Z	"Zabolo"
B	Bamboo
R	River
Rm	River margins
S	Shores
M	Marshes
A	Aguajales
O	Overhead

Foraging Position

T	Terrestrial
U	Undergrowth
Sc	Subcanopy
C	Canopy
W	Water
A	Aerial

Sociality

S	Solitary or in pairs
G	Gregarious
M	Mixed-species flocks
A	Army ant followers

Abundance

C	Common
F	Fairly common
U	Uncommon
R	Rare
(M)	Migrant
(Mn)	Migrant from north
(Ms)	Migrant from south
(Mm)	Migrant from Andes

Evidence

sp	Specimen
t	Tape
si	Species ID by sight
ph	Photo

	Habitats	Foraging	Sociality	Abundance	Evidence
THRAUPINAE (34)					
Cissopis leveriana	Z,Fe	Sc,C	S,M	F	ph,t
Lamprospiza melanoleuca	Fh,Ft	C	M	F	t
Thlypopsis sordida	W,Fe	U	M	R(Ms?)	t
Hemithraupis guira	Ft	C	M	R	si
H. flavicollis	Ft,Fh	C	M	F	t?
Nemosia pileata	W	C	M	R(M?)	si
Lanio versicolor	Ft,Fh	Sc,C	M	F	ph,t
Tachyphonus luctuosus	Z,Ft,Fh	Sc,C	M	C	t?
Habia rubica	Fh,Ft	U	G,M	C	t
Ramphocleus nigrogularis	Ft	Sc,C	G,M	C	t
R. carbo	Fe,Z,Ft	U,Sc,C	G,M	C	ph
Thraupis episcopus	Z,Fe	Sc,C	S,M	U	si
T. palmarum	Ft,Fe	C	S,M	F	t?
Euphonia laniirostris	Z,Fe	Sc,C	S,M	U	t?
E. musica	Fh	C	S	R(M?)	t
E. chrysopasta	Ft	Sc,C	S,M	F	t?
E. minuta	Ft	C	M	U	si
E. xanthogaster	Ft,Fh	U,Sc,C	M	F	t?
E. rufiventris	Ft, Fh	Sc,C	S,M	C	t?
Chlorophonia cyanea	Ft,Fh	C	M	R(M?)	t?
Tangara mexicana	Ft,Fe	C	M,G	F	ph,t
T. chilensis	Ft,Fh,Fe	Sc,C	G,M	C	t?
T. schrankii	Ft,Fh,Fe	U,Sc,C	M	C	t?
T. xanthogastra	Ft,Fh,Fe	C	M	U	si
T. nigrocincta	Ft,Fh	C	M	F	t?
T. velia	Fh	C	M	U	t?
T. callophrys	Fh	C	M	U	t?
Dacnis lineata	Ft,Fh,Fe	C	M	C	t?
D. flaviventer	Ft	C	S,M	R	si
D. cayana	Ft,Fh	C	M	C	ph
Chlorophanes spiza	Ft,Fh,Fe	C	S,M	U	t?
Cyanerpes caeruleus	Ft,Fh	C	S,G,M	C	t?
Tersina viridis	Z,Rm	C	G	U	ph,t

	Habitats	Foraging	Sociality	Abundance	Evidence
PARULIDAE (2)					
Geothlypis aequinoctialis	M,Fe	U	S	F	t?
Basileuterus rivularis	Ft,Fh,Fsm	T,U	S	F	t
ICTERIDAE (10)					
Scaphidura oryzivora	Ft,Rm,S	T,C	S,G	F	t?
Clypicterus oseryi	Ft	C	S, G	R	ph,t
Psarocolius decumanus	Ft,Fh	C	G,M	F	t
P. angustifrons	Ft,Z	Sc,C	G,M	C	t
P. yuracares	Ft,Fs	C	G,M	F	ph,t
Cacicus cela	Ft,Z	Sc,C	G,M	C	t?
C. haemorrhous	Ft,Fh	Sc,C	S,G,M	U	t?
C. solitarius	Z,B	U,Sc	S	F	t?
Icterus cayanensis	Ft,Fe	C	S,M	F	t
I. icterus	Lm,M,Fe,Z	U,Sc	S	U	t

Habitats

Fh	Upland Forest
Ft	Floodplain forest
Fs	Swamp forest
Fe	Forest edge
Fsm	Forest stream margins
Fo	Forest openings
Z	"Zabolo"
B	Bamboo
R	River
Rm	River margins
S	Shores
M	Marshes
A	Aguajales
O	Overhead

Foraging Position

T	Terrestrial
U	Undergrowth
Sc	Subcanopy
C	Canopy
W	Water
A	Aerial

Sociality

S	Solitary or in pairs
G	Gregarious
M	Mixed-species flocks
A	Army ant followers

Abundance

C	Common
F	Fairly common
U	Uncommon
R	Rare
(M)	Migrant
(Mn)	Migrant from north
(Ms)	Migrant from south
(Mm)	Migrant from Andes

Evidence

sp	Specimen
t	Tape
si	Species ID by sight
ph	Photo

Birds of the Tambopata Reserve (Explorer's Inn Reserve)

Theodore A. Parker, III, Paul K. Donahue and Thomas S. Schulenberg

	Habitats	Foraging	Sociality	Abundance	Evidence
TINAMIDAE (9)					
Tinamus major	Ft	T	S	F	t
T. guttatus	Fh,Ft	T	S	F	t
Crypturellus cinereus	Ft	T	S	C	t
C. soui	Ft	T	S	F	t
C. undulatus	Z,Ft	T	S	C	t
C. variegatus	Fh	T	S	R	t
C. bartletti	Ft	T	S	U	t
*C. parvirostris**	C	T	S	R	si
C. atrocapillus	Fh?	T	S	R	?
PODICIPEDIDAE (1)					
Podiceps dominicus	L	W	S	R	si
PHALACROCORACIDAE (1)					
Phalacrocorax brasiliensis	L,R	W	S,G	U	ph
ANHINGIDAE (1)					
Anhinga anhinga	L,R	W	S	U	ph
ARDEIDAE (15)					
Pilherodias pileatus	S,Lm	W	S	F	ph,t
Ardea cocoi	S,Rm,Lm	W	S	U	ph
Casmerodius albus	S,Lm	W	S	R	ph
Bubulcus ibis	S,C,Lm	T	S,G	U	ph
*Egretta caerulea**	S,Rm	W	S	R	si
E. thula	S,Rm,Lm	W	S	U	ph
Butorides striatus	Lm,M	W	S	U	ph,t
Agamia agami	Lm,Fsm,Fs	W	S	R	ph
*Nycticorax nycticorax**	Lm	W	S	R	si
Cochlearius cochlearius	Lm	W	S	U	ph
Tigrisoma lineatum	Lm,Fsm,Fs	W	S	U	ph,t
Zebrilus undulatus	Fs,Fsm	T,W	S	R	ph
*Ixobrychus involucris**	M	T,W	S	R(Ms)	si
I. exilis	M	T,W	S	R(M?)	t
Botaurus pinnatus	M	T,W	S	R(Ms)	ph

	Habitats	Foraging	Sociality	Abundance	Evidence
CICONIIDAE (2)					
Mycteria americana	S	W	G,S	U	si
Jabiru mycteria	S	W	S	R	si
THRESKIORNITHIDAE (2)					
Mesembrinibis cayennensis	Fs,Lm	T,W	S	U	ph
Ajaia ajaja	S	W	S	R	si
ANHIMIDAE (2)					
Anhima cornuta	M,S	T	S	F	ph,t
Chauna torquata	S	T	S	R(Ms)	ph
ANATIDAE (3)					
Dendrocygna sp.	O	W	G	R	si
Neochen jubata	S	T	S	R	si
Cairina moschata	R,L,Fs	W	S	U	ph
CATHARTIDAE (4)					
Sarcoramphus papa	Ft	T	S	U	si
Coragyps atratus	S,Rm,Ft	T	S,G	U	si
Cathartes aura	S,Rm,Ft	T	S	R	si
Cathartes melambrotus	Z,Ft,Fh	T	S	F	si
ACCIPITRIDAE (27)					
*Chondrohierax uncinatus**	Lm	Sc,U	S	R	si
Gampsonyx swainsonii	Rm,Fe	C,A	S	R	si
Elanoides forficatus	Ft	A	S,G	R	si
Leptodon cayanensis	Ft,Fh	C,Sc	S	U	t
Harpagus bidentatus	Fh	Sc	S	U	t
Ictinia plumbea	Z,Ft	A	S,G	U	si
Rostrhamus sociabilis	M	T	S	R	si
R. hamatus	Lm,M	U,T	S	R	si
Accipiter poliogaster	Ft	Sc,C	S	R	si
A. bicolor	Ft	U,Sc	S	R	si
A. superciliosus	Ft,Fh	C,Sc	S	R	t
A. striatus	O	C,Sc	S	R(Ms)	si
Buteo albonotatus	Ft	C,T	S	R	si
*B. platypterus**	Ft,Fe	C,Sc	S	R(Mn)	si
B. magnirostris	Z,Rm,Fe	T,C	S	F	t
B. brachyurus	Ft	A,C	S	R	si

Habitats

Fh	Upland Forest
Ft	Floodplain forest
Fs	Swamp forest
Fsm	Forest stream margins
Fo	Forest openings
Fe	Forest edges
Z	"Zabolo"
B	Bamboo
W	"Willow Bar"
C	Clearing
R	River
Rm	River margins
S	Shores
L	Lake
Lm	Lake margins
M	Marshes
A	Aguajales
O	Overhead

Foraging Position

T	Terrestrial
U	Undergrowth
Sc	Subcanopy
C	Canopy
W	Water
A	Aerial

Sociality

S	Solitary or in pairs
G	Gregarious
M	Mixed-species flocks
A	Army ant followers

Abundance

C	Common
F	Fairly common
U	Uncommon
R	Rare
(M)	Migrant
(Mn)	Migrant from north
(Ms)	Migrant from south
(Mm)	Migrant from Andes
(W)	Wet season only

Evidence

sp	Specimen
t	Tape
si	Species ID by sight
ph	Photo
*****	recorded ≤ 3 times

	Habitats	Foraging	Sociality	Abundance	Evidence
*Asturina nitida**	Z,Fe	T,C	S	R	si
Leucopternis kuhli	Ft,Fh	C	S	R	t
L. schistacea	Fs,Fsm,Ft	T,Sc	S	U	t
Buteogallus urubitinga	Rm,Lm,S	T,U	S	U	ph,t
Busarellus nigricollis	Fs,Fsm,Lm	T,U	S	R	t
Morphnus guianensis	Ft,Fh	Sc,C	S	R	si
Harpia harpya	Ft,Fh	Sc,C	S	R	ph
Spizastur melanoleucus	Ft	C,A?	S	R	si
Spizaetus ornatus	Ft,Fh	C,T	S	R	t
Spizaetus tyrannus	Z,Ft	Sc,C	S	U	t
Geranospiza caerulescens	Ft,Fs	T,C	S	U	t
PANDIONIDAE (1)					
Pandion haliaetus	L,R	W	S	U(Mn)	si
FALCONIDAE (9)					
Herpetotheres cachinnans	Ft,Fh,	C,T	S	U	t
Micrastur mirandollei	Ft,Fh	Sc,U	S	R	t
M. ruficollis	Ft,Fh	Sc,U	S	U	t
M. gilvicollis	Ft,Fh	Sc,U	S	U	t
Daptrius ater	Rm,S,Z	T,C	S,G	U	t
D. americanus	Ft,Fh	C	G	F	t
Milvago chimachima	Rm	T,C	S	R	si
*Falco deiroleucus**	Rm,Fe	C,A	S	R	si
Falco rufigularis	Rm,C	A	S	F	t
CRACIDAE (4)					
Ortalis guttata	Z,Ft	C	G	U	ph,t
Penelope jacquacu	Ft,Fh	C,T	S,G	F	ph,t
Aburria pipile	Ft,Z	C	S	U	ph,t
Mitu tuberosa	Ft,Fh,Fsm	T	S	R	ph,t
PHASIANIDAE (1)					
Odontophorus stellatus	Ft,Fh	T	G	U	t
OPISTHOCOMIDAE (1)					
Opisthocomus hoazin	Lm	U,C	G	C	ph,t
ARAMIDAE (1)					
Aramus guarauna	Lm,M	W	S	R	t

	Habitats	Foraging	Sociality	Abundance	Evidence
PSOPHIIDAE (1)					
Psophia leucoptera	Fh,Ft	T	G	U	t
RALLIDAE (8)					
Rallus nigricans	M	T,W	S	R	si
Aramides cajanea	Lm,Fsm,Fs	T	S	F	t
Anurolimnas castaneiceps	C	T	S	R	t
Laterallus exilis	M	T,W	S	U	t
L. melanophaius	M	T,W	S	U	t
*Neocrex erythrops**	M	T	S	R	si
Porphyrula martinica	M	T,W	S	U	si
P. flavirostris	M	T,W	S	U(M?)	sp
HELIORNITHIDAE (1)					
Heliornis fulica	L	W	S	U	ph
EURYPYGIDAE (1)					
Eurypyga helias	Fsm,Fs,S	T,W	S	U	t
JACANIDAE (1)					
Jacana jacana	M	T,W	S,G	U	t
CHARADRIIDAE (4)					
Hoploxypterus cayanus	S	T	S	F	si
*Vanellus sp.**	O,C,S	T	S	R	si
Charadrius collaris	S	T	S	U	t
Pluvialis dominica	S	T	S,G	R(Mn)	si
SCOLOPACIDAE (9)					
Tringa solitaria	S	T	S	U(Mn)	si
T. flavipes	S	T	S	R(Mn)	si
T. melanoleuca	S	T	S	R(Mn)	si
Actitis macularia	S	T	S	F(Mn)	t
*Calidris fuscicollis**	S	T	S	R(Mn)	si
C. melanotos	S	T	S,G	R(Mn)	si
*C. himantopus**	S	T	S,G	R(Mn)	si
Tryngites subruficollis	S	T	G	R(Mn)	si
Bartramia longicauda	C,S	T	S	R(Mn)	si
LARIDAE (4)					
*Larus pipixcan**	L	W	S	R(Mn)	si
Phaetusa simplex	R,L	W,A	S	U	t

Habitats

Fh	Upland Forest
Ft	Floodplain forest
Fs	Swamp forest
Fsm	Forest stream margins
Fo	Forest openings
Fe	Forest edges
Z	"Zabolo"
B	Bamboo
W	"Willow Bar"
C	Clearing
R	River
Rm	River margins
S	Shores
L	Lake
Lm	Lake margins
M	Marshes
A	Aguajales
O	Overhead

Foraging Position

T	Terrestrial
U	Undergrowth
Sc	Subcanopy
C	Canopy
W	Water
A	Aerial

Sociality

S	Solitary or in pairs
G	Gregarious
M	Mixed-species flocks
A	Army ant followers

Abundance

C	Common
F	Fairly common
U	Uncommon
R	Rare
(M)	Migrant
(Mn)	Migrant from north
(Ms)	Migrant from south
(Mm)	Migrant from Andes
(W)	Wet season only

Evidence

sp	Specimen
t	Tape
si	Species ID by sight
ph	Photo
*****	recorded ≤ 3 times

	Habitats	Foraging	Sociality	Abundance	Evidence
Sterna superciliaris	R,L	W,A	S	F	t
*Sterna nilotica**	R	W,A	S	R(Mn)	si
RHYNCHOPIDAE (1)					
Rynchops niger	R,L	W	S,G	R	ph,t
COLUMBIDAE (9)					
*Columba speciosa**	O(Rm)	C	S	R	si
C. cayennensis	Lm,Z	U,C	S,G	C	ph,t
C. subvinacea	Ft	C	S	U	t
C. plumbea	Ft,Fh	C	S	C	t
Columbina talpacoti	C,Rm,Lm	T	S,G	U	si
C. picui	C,Rm	T	S,G	U(Ms?)	si
Claravis pretiosa	Z	T	S	R	t
Leptotila rufaxilla	Z,C	T	S	F	ph,t
Geotrygon montana	Ft,Fh	T	S	F	ph,t
PSITTACIDAE (19)					
Ara ararauna	A,Fs,Ft	C	G	U	ph,t
A. macao	Ft,Fh	C	S,G	F	ph,t
A. chloroptera	Ft,Fh	C	S,G	F	ph,t
A. severa	Z,Ft	C	S,G	F	ph,t
A. manilata	Ft	C	G	R	ph,t
A. couloni	Ft	C	S,G	R	t
Aratinga leucophthalmus	Z,Ft	C	G	U	ph,t
A. weddellii	Z,Ft	C	G	C	ph,t
Pyrrhura rupicola	Ft,Fh	C	G	F	t
Forpus sclateri	Ft,Fh	C	S	U	t
Brotogeris cyanoptera	Z,Ft,Fh	C	G	C	ph,t
*B. sanctithomae**	Z	C	G	R	si
Touit huetii	?(O)	C	G	R	t
Nannopsittaca dachilleae	Ft	C	S	R	t
Pionites leucogaster	Ft,Fh	C	G	F	t
Pionopsitta barrabandi	Fh,Ft	Sc,C	S,G	F	t
Pionus menstruus	Ft	C	S,G	C	t
Amazona ochrocephala	Ft,Z	C	S	U	t
A. farinosa	Ft,Fh	C	S,G	C	t

	Habitats	Foraging	Sociality	Abundance	Evidence
CUCULIDAE (13)					
*Coccyzus cinereus**	Ft,W,Z	C	M,S	R(Ms)	si
*C. erythophthalmus**	Fh	C	M,S	R(Mn)	si
C. americanus	Ft,Fh	C,Sc	M,S	U(Mn)	si
C. melacoryphus	Z,Ft	C	S,M	U(Ms)	si
Piaya cayana	Z,Ft,Fh	Sc,C	S,M	F	t
P. melanogaster	Ft,Fh	C	S,M	U	t
P. minuta	Lm	U,Sc	S	U	t
Crotophaga major	Rm,Lm,Fs	U,Sc	G	F(W)	t
C. ani	C,Lm	U,T	G	F	t
Tapera naevia	C	T,U	S	R	t
Dromococcyx phasianellus	Ft	T,U	S	R	t
D. pavoninus	Ft,B	T,U	S	R	t
Neomorphus geoffroyi	Fh	T	S	R	si
STRIGIDAE (8)					
Otus choliba	Z	Sc,C	S	R	t
O. watsonii	Ft,Fh	U,C	S	C	t
Lophostrix cristata	Ft,Fh	C	S	U	t
Pulsatrix perspicillata	Ft	C,T	S	U	t
Glaucidium hardyi	Ft,Fh	C	S	F	t
G. brasilianum	Z	C,Sc	S	U	t
Ciccaba virgata	Ft,Fh	C	S	U	sp
C. huhula	Ft,Fh?	C	S	R	?
NYCTIBIIDAE (4)					
Nyctibius grandis	Ft	C,A	S	U	t
N. aethereus	Ft	C,A	S	R	t
N. griseus	Z,Ft,Fe	C,A	S	U	t
*N. bracteatus**	Ft	?	S	R	si
CAPRIMULGIDAE (8)					
Lurocalis semitorquatus	O(Ft)	A	S	R	t
Chordeiles rupestris	S	A	G	C	si
C. minor	Fe	A	G	R(Mn)	si
Nyctidromus albicollis	Z,C	A	S	F	t
Nyctiphrynus ocellatus	Ft,B	A,Sc	S	U	t
Hydropsalis climacocerca	Rm,S	A	S	F	t

Habitats

Fh	Upland Forest
Ft	Floodplain forest
Fs	Swamp forest
Fsm	Forest stream margins
Fo	Forest openings
Fe	Forest edges
Z	"Zabolo"
B	Bamboo
W	"Willow Bar"
C	Clearing
R	River
Rm	River margins
S	Shores
L	Lake
Lm	Lake margins
M	Marshes
A	Aguajales
O	Overhead

Foraging Position

T	Terrestrial
U	Undergrowth
Sc	Subcanopy
C	Canopy
W	Water
A	Aerial

Sociality

S	Solitary or in pairs
G	Gregarious
M	Mixed-species flocks
A	Army ant followers

Abundance

C	Common
F	Fairly common
U	Uncommon
R	Rare
(M)	Migrant
(Mn)	Migrant from north
(Ms)	Migrant from south
(Mm)	Migrant from Andes
(W)	Wet season only

Evidence

sp	Specimen
t	Tape
si	Species ID by sight
ph	Photo
*	recorded ≤ 3 times

	Habitats	Foraging	Sociality	Abundance	Evidence
Caprimulgus parvulus	Ft	A,Sc	S	R(Ms)	t
C. sericocaudatus	Ft,Fh	A,C	S	R(Ms?)	t
APODIDAE (8)					
Streptoprocne zonaris	O(Ft,Fh)	A	G	F	si
Cypseloides sp.	O(Ft)	A	G	R(M?)	si
Chaetura chapmani	O(Ft)	A	G,M	R(M?)	si
Chaetura cinereiventris	O(Ft)	A	G,M	C(M?)	t
C. egregia	O(Ft)	A	G,M	U(Ms?)	t?
C. brachyura	O(Fe)	A	G,M	U	t
Panyptila cayennensis	O(Ft)	A	S,M	U	t
Reinarda squamata	O(A,L,R)	A	S,G	C	t
TROCHILIDAE (18)					
Glaucis hirsuta	Z,Ft	U,Sc	S	U	t
Threnetes leucurus	Z,Ft	U	S	R	sp,t
Phaethornis philippi	Fh	U	S	F	t
P. hispidus	Z,Ft	U	S	F	sp,t
P. ruber	Ft,Fh,B	U	S	C	t
Campylopterus largipennis	Ft,Z	U,Sc	S	R	si
Florisuga mellivora	Ft,Fh	C	S	U	si
Anthracothorax nigricollis	Rm,Lm	C	S	U	si
Lophornis chalybea	Ft,W	C	S	R	si
Popelairia sp.	C	C	S	R	si
Thalurania furcata	Ft,Fh	U,C	S	F	t
Hylocharis cyanus	Ft	C	S	F	t
Chrysuronia oenone	C,W	C	S	R	si
Amazilia lactea	Z,Ft	C	S	R	si
Polyplancta aurescens	Ft	Sc,C	S	U	t
Heliothryx aurita	Ft,Fh	Sc,C	S	U	si
Heliomaster longirostris	Z,Rm	C	S	R	si
Calliphlox amethystina	W	C	S	R	si
TROGONIDAE (6)					
Pharomachrus pavoninus	Fh,Ft	C	S	U	t
Trogon melanurus	Ft,Fh	C	S	C	t
T. viridis	Ft,Fh	C,Sc	S,M	C	t
T. collaris	Ft,Fh	Sc,U	S,M	F	t

	Habitats	Foraging	Sociality	Abundance	Evidence
T. curucui	Ft,Z	C	S,M	U	t
T. violaceus	Ft,Fh	C	S,M	U	t
ALCEDINIDAE (5)					
Ceryle torquata	Rm,Lm	W	S	F	t
Chloroceryle amazona	Rm,Lm	W	S	F	t
C. americana	Fsm,Rm,Lm	W	S	U	si
C. inda	Fsm,Lm	W	S	U	si
C. aenea	Fsm,Lm	W	S	U	si
MOMOTIDAE (3)					
Electron platyrhynchum	Ft,Fh	C,Sc	S	C	t
Barypthengus martii	Ft	Sc,C	S	F	t
Momotus momota	Ft,Fh	Sc,C	S	U	t
GALBULIDAE (4)					
Brachygalba albogularis	Z	C,A	S	R	ph,t
Galbula cyanescens	Ft,Fo	U,Sc	S,M	F	sp,t
G. dea	Fh	C	S,M	U	t
Jacamerops aurea	Ft,Fh	C,Sc	S	U	t
BUCCONIDAE (10)					
Notharchus macrorhynchos	Ft,Fh	C	S	U	t
N. ordii	Ft	C	S	R	ph
Bucco macrodactylus	Z	Sc,C	S	R	si
Nystalus striolatus	Ft,Fh	C,	S	U	t
Malacoptila semicincta	Ft,Fh	U	S	U	t
Nonnula ruficapilla	Ft,Z,B	U,Sc	S	U	sp,t
Monasa nigrifrons	Ft,Z,	C,Sc	G,M	C	t
M. morphoeus	Ft,Fh	C,Sc	G,M	C	t
M. flavirostris	Ft	C	S,M	R	t
Chelidoptera tenebrosa	Rm,Z,Ft	C,A	S	F	t
CAPITONIDAE (3)					
Capito niger	Ft,Fh	C	S,M	C	t
Eubucco richardsoni	Ft	C	S,M	F	t
E. tucinkae	Z,Ft	C	S,M	R	t
RAMPHASTIDAE (8)					
Aulacorhynchus prasinus	Ft	C	S	U	t
Pteroglossus castanotis	Ft,Z	C	G	F	t

Habitats

Fh	Upland Forest
Ft	Floodplain forest
Fs	Swamp forest
Fsm	Forest stream margins
Fo	Forest openings
Fe	Forest edges
Z	"Zabolo"
B	Bamboo
W	"Willow Bar"
C	Clearing
R	River
Rm	River margins
S	Shores
L	Lake
Lm	Lake margins
M	Marshes
A	Aguajales
O	Overhead

Foraging Position

T	Terrestrial
U	Undergrowth
Sc	Subcanopy
C	Canopy
W	Water
A	Aerial

Sociality

S	Solitary or in pairs
G	Gregarious
M	Mixed-species flocks
A	Army ant followers

Abundance

C	Common
F	Fairly common
U	Uncommon
R	Rare
(M)	Migrant
(Mn)	Migrant from north
(Ms)	Migrant from south
(Mm)	Migrant from Andes
(W)	Wet season only

Evidence

sp	Specimen
t	Tape
si	Species ID by sight
ph	Photo
*	recorded ≤ 3 times

	Habitats	Foraging	Sociality	Abundance	Evidence
P. inscriptus	Ft	C	G	R	t
P. mariae	Ft,Fh	C	G	F	t
Pteroglossus beauharnaesii	Ft,Fh	C	G	U	t
Selenidera reinwardtii	Ft,Fh	C,Sc	S	F	t
Ramphastos culminatus	Ft,Fh	C	S,G	C	t
R. cuvieri	Ft,Fh	C	S,G	C	t
PICIDAE (16)					
Picumnus rufiventris*	Z	Sc	S,M	R	si
P. aurifrons	Ft,Fh	C	S,M	U	sp
Chrysoptilus punctigula	Z	Sc,C	S	R	t
Piculus leucolaemus	Ft	C	M	U	t
P. chrysochloros	Ft,Fh	C,Sc	S,M	U	t
Celeus elegans	Ft,Fs	C,Sc	S,M	R	t
C. grammicus	Fh,Ft	C	S,M	F	t
C. flavus	Ft	C,Sc	S,G	U	t
C. spectabilis	Z,Ft,B	U,Sc	S,M	U	ph,t
C. torquatus	Ft,Fh	C,Sc	S	R	t
Dryocopus lineatus	Z,Ft	Sc,C	S	F	t
Melanerpes cruentatus	Ft,Fh,Fo	C	S,M	C	t
Veniliornis passerinus	Z	Sc,C	S,M	F	t
V. affinis	Ft,Fh	C	M	F	t
Campephilus melanoleucus	Ft,Z	C,Sc	S	F	t
C. rubricollis	Ft,Fh	U,Sc	S	F	t
DENDROCOLAPTIDAE (16)					
Dendrocincla fuliginosa	Ft	U,Sc	S,M,A	F	t
D. merula	Ft,Fh	U,Sc	S,A	U	t
Deconychura longicauda	Ft,Fh	Sc	S,M	U	t
Sittasomus griseicapillus	Ft	U,C	M	F	t
Glyphorynchus spirurus	Ft,Fh	U,Sc	S,M	C	sp
Nasica longirostris	Lm,Fs,Ft	Sc,C	S	U	t
Dendrexetastes rufigula	Ft,Fh	C,Sc	S,M	F	t
Xiphocolaptes promeropirhynchus	Fh	U,C	S,M	R	t
Dendrocolaptes certhia	Ft,Fh	Sc,C	S,M,A	F	t
D. picumnus	Ft,Fh	U,Sc	S,A	U	t
Xiphorhynchus picus	Z	Sc	S	R	t

	Habitats	Foraging	Sociality	Abundance	Evidence
X. obsoletus	A?	U,Sc	S,M	R	t
X. spixii	Ft,Fh	U,Sc	M	C	sp,t
X. guttatus	Ft,Fh	Sc,C	S,M	C	t
Lepidocolaptes albolineatus	Ft,Fh	C	M	F	t
Campylorhamphus trochilirostris	Ft,B	U	S,M	U	sp
FURNARIIDAE (28)					
Furnarius leucopus	Z,Rm	T	S	F	t
Synallaxis cabanisi	B	U	S	R	t
S. albescens	Rm,C	U	S	R(Ms?)	sp
S. albigularis	Rm,C	U	S	F	t
S. gujanensis	Z	U	S	F	t
S. rutilans	Ft,Fh	T,U	S	F	t
Cranioleuca gutturata	Ft	C,Sc	M	U	t
Thripophaga fusciceps	Ft	U,Sc	S,M	R	sp
Berlepschia rikeri	A	C	S	R	t
Hyloctistes subulatus	Ft	Sc,C	S,M	U	sp,t
Ancistrops strigilatus	Ft,Fh	Sc,C	M	C	sp,t
Simoxenops ucayale	Ft,B,Z	U	M,S	U	sp
Philydor erythrocercus	Fh,Ft	Sc	M	F	t
P. pyrrhodes	Ft,Fh	Sc,U	S,M	U	t
P. rufus	Z	C	S,M	U	t
P. erythropterus	Ft,Fh	C	M	F	t
P. ruficaudatus	Ft	Sc,C	M	R	t
Automolus infuscatus	Ft,Fh	U	M	C	sp,t
A. dorsalis	Ft,Z,B	Sc	S,M	R	t
A. rubiginosus	Ft,B	U	S	R	t
A. ochrolaemus	Ft,Fh	U	S,M	F	ph,t
A. rufipileatus	Z,Ft	U	S,M	F	t
A. melanopezus	Ft,B	U,Sc	S,M	U	sp,t
Xenops milleri	Ft,Fh	C	M	F	t
X. rutilans	Ft,Fh	C	M	R	t
X. minutus	Ft,Fh	U,Sc	M	F	sp,t
*Sclerurus mexicanus**	Ft	T	S	R	si
S. caudacutus	Ft,Fh	T	S	F	t

Habitats

Fh	Upland Forest
Ft	Floodplain forest
Fs	Swamp forest
Fsm	Forest stream margins
Fo	Forest openings
Fe	Forest edges
Z	"Zabolo"
B	Bamboo
W	"Willow Bar"
C	Clearing
R	River
Rm	River margins
S	Shores
L	Lake
Lm	Lake margins
M	Marshes
A	Aguajales
O	Overhead

Foraging Position

T	Terrestrial
U	Undergrowth
Sc	Subcanopy
C	Canopy
W	Water
A	Aerial

Sociality

S	Solitary or in pairs
G	Gregarious
M	Mixed-species flocks
A	Army ant followers

Abundance

C	Common
F	Fairly common
U	Uncommon
R	Rare
(M)	Migrant
(Mn)	Migrant from north
(Ms)	Migrant from south
(Mm)	Migrant from Andes
(W)	Wet season only

Evidence

sp	Specimen
t	Tape
si	Species ID by sight
ph	Photo
*****	recorded ≤ 3 times

	Habitats	Foraging	Sociality	Abundance	Evidence
FORMICARIIDAE (52)					
Cymbilaimus lineatus	Ft,Fh	Sc,C	S,M	F	t
C. sanctaemariae	Ft,B	Sc,C	S,M	F	ph,t
Frederickena unduligera	Fh	U	S	R	t
Taraba major	Z,Lm	U	S	U	t
Thamnophilus doliatus	Rm,C	U	S,M	U	t
T. aethiops	Ft,Fh	U	S	F	t
T. schistaceus	Ft,Fh	Sc	M	C	t
T. amazonicus	Fs,Lm	Sc	S	R	t
Pygiptila stellaris	Ft,Fh	C,Sc	M	C	t
Thamnomanes ardesiacus	Ft,Fh	U	M	F	t
T. schistogynus	Ft	U,Sc	M	C	sp,t
Myrmotherula brachyura	Ft,Fh	C	M	C	t
M. sclateri	Ft,Fh	C	M	F	t
M. surinamensis	Lm	U,Sc	S,M	F	t
M. longicauda	Z	U	S	R	t?
M. hauxwelli	Ft,Fh	U	S,M	F	sp,t
M. leucophthalma	Ft,Fh	U	M	F	sp,t
M. ornata	Ft,B	U,Sc	M	F	sp,t
M. axillaris	Ft,Fh	U,Sc	M	C	sp,t
M. iheringi	Ft,B	U	M	U	sp,t
M. longipennis	Fh	U,Sc	M	F	t
M. menetriesii	Ft,Fh	Sc,C	M	C	t
Dichrozona cincta	Fh	T	S	R	t
Microrhopias quixensis	Ft,B	Sc	M	R	t
Drymophila devillei	Ft,B	Sc	S,M	F	t
Terenura humeralis	Ft	C	M	U	t
Cercomacra cinerascens	Ft,Fh	C,Sc	S,M	C	t
C. nigrescens	Z,B	U	S	U	ph,t
C. serva	Fh,Ft,Fo	U	S	R	ph,t
C. manu	Z,B	U	S	R	t
Myrmoborus leucophrys	Ft	U	S	F	sp,t
M. myotherinus	Ft,Fh	U	S,A	C	t
Hypocnemis cantator	Ft,B	U	S,M	C	sp,t
Hypocnemoides maculicauda	Fs,Lm,Fsm	U,T	S	F	sp

CONSERVATION INTERNATIONAL

Rapid Assessment Program

	Habitats	Foraging	Sociality	Abundance	Evidence
Percnostola lophotes	Z,B	T,U	S	F	sp,t
Sclateria naevia	Fs,Fsm,Lm	T	S	U	t
Myrmeciza hemimelaena	Ft,Fh	U,T	S	C	t
M. hyperythra	Fs,Lm	T,U	S	F	t
M. goeldii	Ft,B	U	S,A	U	sp,t
M. atrothorax	Z	U	S	U	t
Gymnopithys salvini	Ft,Fh	U	A	F	sp
*Rhegmatorhina melanosticta**	Fh	U	A	R	si
H. punctulata	Ft	U	S	R	ph
H. poecilinota	Ft,Fh	U	S,A	U	ph,t
Phlegopsis nigromaculata	Ft	T,U	S,A	U	ph,t
Chamaeza nobilis	Ft	T	S	R	t
Formicarius colma	Fh	T	S	F	ph,t
F. analis	Ft	T	S,A	C	ph,t
F. rufifrons	Ft	T	S	R	t
Hylopezus berlepschi	Z	T	S	R	t
Myrmothera campanisona	Ft,Fh	T	S	F	t
Conopophaga peruviana	Ft	U,T	S	R	t
COTINGIDAE (7)					
Iodopleura isabellae	Ft	C	S	U	t
*Porphyrolaema porphyrolaema**	Ft	C	M	R	si
Lipaugus vociferans	Ft,Fh	Sc,C	S	C	t
Cotinga maynana	Ft,Z	C	S	F	t
C. cayana	Fh	C	S	U	si
Gymnoderus foetidus	Ft	C	S	F	si
Querula purpurata	Fh,Ft	C	G	C	t
PIPRIDAE (10)					
Schiffornis major	Fs,Lm	U	S	U	sp
S. turdinus	Fh	U	S	U	t
Piprites chloris	Ft,Fh	C	M	F	t
Xenopipo atronitens	Z	U	S	R	sp
Tyranneutes stolzmanni	Ft,Fh	Sc	S	C	t
Machaeropterus pyrocephalus	Ft,B	Sc,C	S	F	sp,t
Pipra coronata	Fh	U,Sc	S	U	t
P. fasciicauda	Ft	U,Sc	S	F	sp,t

Habitats

Fh	Upland Forest
Ft	Floodplain forest
Fs	Swamp forest
Fsm	Forest stream margins
Fo	Forest openings
Fe	Forest edges
Z	"Zabolo"
B	Bamboo
W	"Willow Bar"
C	Clearing
R	River
Rm	River margins
S	Shores
L	Lake
Lm	Lake margins
M	Marshes
A	Aguajales
O	Overhead

Foraging Position

T	Terrestrial
U	Undergrowth
Sc	Subcanopy
C	Canopy
W	Water
A	Aerial

Sociality

S	Solitary or in pairs
G	Gregarious
M	Mixed-species flocks
A	Army ant followers

Abundance

C	Common
F	Fairly common
U	Uncommon
R	Rare
(M)	Migrant
(Mn)	Migrant from north
(Ms)	Migrant from south
(Mm)	Migrant from Andes
(W)	Wet season only

Evidence

sp	Specimen
t	Tape
si	Species ID by sight
ph	Photo
*****	recorded ≤ 3 times

	Habitats	Foraging	Sociality	Abundance	Evidence
P. rubrocapilla	Fh	U,Sc	S	U	sp,t
P. chloromeros	Ft,Fh	U,Sc	S	F	t
TYRANNIDAE (90)					
Zimmerius gracilipes	Ft,Fh	C	S,M	F	t
Ornithion inerme	Ft,Fh	C	S,M	F	sp,t
Camptostoma obsoletum	Z	C	S,M	U	t
Sublegatus modestus	W,C	Sc,C	S,M	R(M?)	si
S. obscurior	Z,C	Sc,C	S,M	R	t
Phaeomyias murina	C	C	S,M	U	t
Tyrannulus elatus	Ft,Fh	C,Sc	S,M	F	t
Myiopagis gaimardii	Ft,Fh	C	M	C	sp,t
M. caniceps	Ft	C	M	R	t
M. viridicata	Ft,Fh,Z	Sc,C	M	U(Ms)	t
Elaenia spectabilis	Z	C,Sc	S,M	U(Ms)	sp
E. parvirostris	Z	C,Sc	S,M	R(Ms)	si
*E. strepera**	B,Ft	U,Sc	S?	R(Ms)	si
Inezia inornata	Z	C	G,M	F(Ms)	sp,t
Pseudocolopteryx acutipennis	W,M	U	S	R(Ms?)	sp
Euscarthmus meloryphus	Z	U	S	R(Ms)	t
Mionectes olivaceus	Ft,Fh	C,Sc	M	R	si
M. oleagineus	Ft,Fh	U,C	M	F	sp,t
M. macconnelli	Ft,Fh	U,Sc	M	R	t?
Leptopogon amaurocephalus	Ft	U,Sc	M,S	F	t
Corythopis torquata	Ft,Fh	T	S	F	t
Myiornis ecaudatus	Ft	C,Sc	S	F	t
Poecilotriccus albifacies	B,Ft	U,Sc	M,S	U	sp
Hemitriccus flammulatus	Ft,B	Sc	S	F	sp,t
H. zosterops	Ft,Fh	Sc	S	F	sp,t
H. iohannis	Z	Sc,C	S	U	t
H. striaticollis	Z	U	S	R(Ms)	sp
Todirostrum latirostre	Z	U	S	F	ph,t
T. maculatum	Z	C,Sc	S,M	F	t
T. chrysocrotaphum	Ft,Fh	C	S,M	F	t
Ramphotrigon megacephala	Z,Ft,B	Sc,U	S,M	F	sp,t
R. fuscicauda	Ft,B	Sc	S	U	ph,t

	Habitats	Foraging	Sociality	Abundance	Evidence
R. ruficauda	Ft,Fh	Sc	S,M	F	t
Rhynchocyclus olivaceus	Ft	U,Sc	M	U	ph,t
Tolmomyias assimilis	Ft,Fh	C	M	F	t
T. poliocephalus	Ft,Z	C	S,M	F	t
T. flaviventris	Z	C,Sc	S,M	F	t
Platyrinchus coronatus	Ft	Sc,U	S	F	sp
P. platyrhynchos	Fh	U,Sc	S	R	t
Onychorhynchus coronatus	Ft,Fh,Fsm	U,Sc	S,M	U	t
Terenotriccus erythrurus	Ft,Fh	U,Sc	S,M	U	t
Myiophobus fasciatus	C,Z	U	S,M	U	t
Contopus borealis	Rm	C	S	R(Mn)	si
C. virens	C,Z	C,A	S	U(Mn)	si
C. cinereus	Z	C,A	S	R(M?)	si
Empidonax alnorum	C,Z	U	S	R(Mn)	sp
Lathrotriccus euleri	Ft,Z,B	U,Sc	S	F	t
Cnemotriccus fuscatus	Z	U,Sc	S	U	t
Pyrocephalus rubinus	Rm,C	C,A	S	F(Ms)	si
Ochthoeca littoralis	Rm,S	T,A	S	F	t
Muscisaxicola fluviatilis	Rm,S	T	S	U	si
Knipolegus sp.*	Lm,M	U	S	R(Ms?)	si
*Hymenops perspicillata**	M	U	S	R(Ms)	si
Fluvicola pica	Lm,M	U,T	S	R(Ms?)	si
Colonia colonus	Rm,Z	C	S	R(Mm?)	t
Satrapa icterophrys	Rm	Sc	S,M	R(Ms)	sp
Attila cinnamomeus	Ft	Sc	S	R	t
A. bolivianus	Ft	Sc,C	S	U	t
A. spadiceus	Ft,Fh	Sc,C	S,M	F	t
Casiornis rufa	Ft	C	M	R(Ms)	sp
Rhytipterna simplex	Ft,Fh	Sc,C	S,M	C	t
Laniocera hypopyrra	Ft,Fh	Sc,C	S,M	F	t
Sirystes sibilator	Ft,Fh	C,Sc	S,M	F	t
Myiarchus tuberculifer	Ft	C	S	R(N?)	t
M. swainsoni	Z,Ft	C	M	U(Ms)	t?
M. ferox	Z,C	Sc,C	S,M	F	sp,t
M. tyrannulus	Ft,Fh	C	S,M	F(Ms)	t

Habitats

Fh	Upland Forest
Ft	Floodplain forest
Fs	Swamp forest
Fsm	Forest stream margins
Fo	Forest openings
Fe	Forest edges
Z	"Zabolo"
B	Bamboo
W	"Willow Bar"
C	Clearing
R	River
Rm	River margins
S	Shores
L	Lake
Lm	Lake margins
M	Marshes
A	Aguajales
O	Overhead

Foraging Position

T	Terrestrial
U	Undergrowth
Sc	Subcanopy
C	Canopy
W	Water
A	Aerial

Sociality

S	Solitary or in pairs
G	Gregarious
M	Mixed-species flocks
A	Army ant followers

Abundance

C	Common
F	Fairly common
U	Uncommon
R	Rare
(M)	Migrant
(Mn)	Migrant from north
(Ms)	Migrant from south
(Mm)	Migrant from Andes
(W)	Wet season only

Evidence

sp	Specimen
t	Tape
si	Species ID by sight
ph	Photo
*	recorded ≤ 3 times

	Habitats	Foraging	Sociality	Abundance	Evidence
Pitangus lictor	M	U	S	R	t
P. sulphuratus	Rm,C	U,Sc	S	F	t
Megarhynchus pitangua	C,Z	Sc,C	S	F	t
Myiozetetes cayanensis	Lm,M	U	S	F	t
M. similis	C,Rm	U,Sc,C	S,G	C	t
M. granadensis	Z,C	Sc,C	S,G	F	t
M. luteiventris	Ft	C	S,G	U	t
Myiodynastes maculatus	Ft	Sc,C	M	U(Ms)	si
M. luteiventris	Ft,Z,W	C	S,M	U(Mn)	si
Legatus leucophaius	Ft,Fo,Lm	C	S	F	ph,t
Empidonomus varius	Ft	C	M	R(Ms)	si
E. aurantioatrocristatus	Ft,Fh,Fo	C	S,M	F(Ms)	si
Tyrannopsis sulphurea	Lm,A	C	S	U	t
Tyrannus melancholicus	Rm,C	C,A	S	F	t
T. savanna	Rm,Lm	C	G	U(Ms?)	si
T. tyrannus	Ft,Rm,Lm	C	G	C(Mn)	si
Pachyramphus castaneus	Z	C	M	R	t?
P. polychopterus	Z,C	C	S,M	F	sp,t
P. marginatus	Ft,Fh	C	M	F	t
P. minor	Ft,Fh	C	M	U	t
P. validus	Fh	C	M	R	si
Tityra cayana	Ft	C	S	F	si
T. semifasciata	Ft,Fh	C	S	U	t
T. inquisitor	Ft	C	S	U	si
HIRUNDINIDAE (13)					
Tachycineta albiventer	R	A	S,G	C	t
*T. leucorrhoa**	C,O	A	G	R(Ms)	si
Phaeprogne tapera	R	A	G,S	U	t
Progne chalybea	R	A	S,G	R	si
*P. modesta**	R,L	A	G	R(Ms)	si
Notiochelidon cyanoleuca	R	A	G	R(Ms?)	si
Atticora fasciata	R,C	A	G	C	t
*Neochelidon tibialis**	Fh	A	U	R	si
*Alopochelidon fucata**	R,L	A	G	R(Ms)	si
Stelgidopteryx ruficollis	R,C	A	G	C	t

	Habitats	Foraging	Sociality	Abundance	Evidence
Riparia riparia	L	A	G	R(Mn)	si
Hirundo rustica	R,C,L	A	G	F(Mn)	si
Petrochelidon pyrrhonota	L	A	G	R(Mn)	si
CORVIDAE (1)					
Cyanocorax violaceus	Z,Ft	Sc,C	G,M	C	t
TROGLODYTIDAE (7)					
Campylorhynchus turdinus	C,Z	Sc,C	S,M	F	t
Thryothorus genibarbis	Z,Ft	U	S	C	sp,t
T. leucotis	C	U	S	U	sp,t
Troglodytes aedon	C	U	S	U	t
Microcerculus marginatus	Ft,Fh	T	S	F	t
Cyphorhinus arada	Ft, Fh	T	S,M	F	t
Donacobius atricapillus	M,C	U	S	F	t
TURDIDAE (8)					
*Catharus fuscescens**	Ft	T,U	S	R(Mn)	si
*C. minimus**	Ft	T,U	S	R(Mn)	si
*C. ustulatus**	Ft,Z	T,U	S	R(Mn)	si
Turdus amaurochalinus	Z,C,Ft	T,Sc,C	S	F(Ms?)	sp,t
T. ignobilis	Z,W,C	T,Sc,C	S,M	U	si
T. lawrencii	Ft	T,Sc,C	S	R	t
T. hauxwelli	Ft	T,Sc	S	U	t
T. albicollis	Ft,Fh	T,Sc	S	F	sp,t
VIREONIDAE (5)					
Cyclarhis gujanensis	Z	C	S,M	R	t
Vireo olivaceus	Ft,Fh,Z	C	M	C(Ms)	t
Hylophilus thoracicus	Ft	C	M	U	t
H. hypoxanthus	Ft,Fh	C	M	C	t
H. ochraceiceps	Fh	U	M	R	t
EMBERIZINAE (13)					
Myospiza aurifrons	S,C	T,U	S	C	t
Volatinia jacarina	C,M	T,U	S,G	R	si
Sporophila schistacea	Ft,B	U,Sc	S	R	sp
S. lineola	M,W,C	U	G,M	U(M?)	si
*S. luctuosa**	W	U	G,M	R(Mm?)	si
S. nigricollis	W,C	U	G,M	R(Mm?)	si

Habitats

Fh	Upland Forest
Ft	Floodplain forest
Fs	Swamp forest
Fsm	Forest stream margins
Fo	Forest openings
Fe	Forest edges
Z	"Zabolo"
B	Bamboo
W	"Willow Bar"
C	Clearing
R	River
Rm	River margins
S	Shores
L	Lake
Lm	Lake margins
M	Marshes
A	Aguajales
O	Overhead

Foraging Position

T	Terrestrial
U	Undergrowth
Sc	Subcanopy
C	Canopy
W	Water
A	Aerial

Sociality

S	Solitary or in pairs
G	Gregarious
M	Mixed-species flocks
A	Army ant followers

Abundance

C	Common
F	Fairly common
U	Uncommon
R	Rare
(M)	Migrant
(Mn)	Migrant from north
(Ms)	Migrant from south
(Mm)	Migrant from Andes
(W)	Wet season only

Evidence

sp	Specimen
t	Tape
si	Species ID by sight
ph	Photo
*	recorded ≤ 3 times

	Habitats	Foraging	Sociality	Abundance	Evidence
S. caerulescens	M,C	T,U	G,M	C(Ms)	sp
*S. leucoptera**	W,M	U	M	R(Ms)	si
S. castaneiventris	C	U,C	S,M	U	si
Oryzoborus maximiliani	M	U	S	R	t
O. angolensis	C,M	U	S	R	si
Arremon taciturnus	Ft,Fh	T,U	S	F	t
Paroaria gularis	Rm	U,Sc,C	S,M	F	sp
CARDINALINAE (7)					
*Pheucticus aureoventris**	C	Sc	S	R(Ms?)	si
*P. ludovicianus**	Ft	C	M	R(Mn)	si
Caryothraustes humeralis	Ft,Fh	C	M	R	si
Pitylus grossus	Ft	Sc	S,M	C	ph,t
Saltator maximus	Z,Ft,Fh	Sc,C	S,M	F	t
S. coerulescens	Z,C	U,Sc,C	S	F	t
Cyanocompsa cyanoides	Ft	U	S	F	t
THRAUPINAE (40)					
Conothraupis speculigera	Z	U,Sc	M	R(M?)	si
Lamprospiza melanoleuca	Fh,Ft	C	M	F	si
Cissopis leveriana	Z,C	Sc,C	S,M	F	t
Thlypopsis sordida	W,C,B	U	M	R(Ms?)	si
Hemithraupis guira	Ft	C	M	R	si
H. flavicollis	Ft,Fh	C	M	F	t
*Nemosia pileata**	W	C	M	R(M?)	si
Lanio versicolor	Ft,Fh	Sc,C	M	F	t
Tachyphonus cristatus	Fh	C	M	U	si
T. luctuosus	Z,Ft,Fh	Sc,C	M	C	t
Habia rubica	Fh,Ft	U	G,M	C	t
*Piranga rubra**	Ft	C	M	R(Mn)	si
P. olivacea	Ft	C	M	R(Mn)	si
Ramphocelus nigrogularis	Ft	Sc,C	G,M	C	t
R. carbo	C,Z,Ft	U,Sc,C	G,M	C	t
Thraupis episcopus	Z,C	Sc,C	S,M	U	si
T. palmarum	Ft,C	C	S,M	F	t
*Pipraeidea melanonota**	C,W	C	M	R(Mm?)	si
Euphonia laniirostris	Z,C	Sc,C	S,M	U	t

	Habitats	Foraging	Sociality	Abundance	Evidence
*E. musica**	Fh	C	S	R(M?)	si
E. chrysopasta	Ft	Sc,C	S,M	F	t
E. minuta	Ft	C	M	U	t?
E. xanthogaster	Ft,Fh	U,Sc,C	M	F	t
E. rufiventris	Ft, Fh	Sc,C	S,M	C	t
Chlorophonia cyanea	Ft,Fh	C	M	R(M?)	t
Tangara mexicana	Ft	C	M,G	F	t
T. chilensis	Ft,Fh	Sc,C	G,M	C	t
T. schrankii	Ft,Fh	U,Sc,C	M	C	t
T. xanthogastra	Ft,Fh	C	M	R	si
*T. gyrola**	Ft	C	G	R	si
T. nigrocincta	Ft,Fh	C	M	F	t
T. velia	Fh	C	M	U	t
T. callophrys	Fh	C	M	U	t
Dacnis lineata	Ft,Fh	C	M	C	t
D. flaviventer	Ft	C	S,M	U	t
D. cayana	Ft,Fh	C	M	C	si
Chlorophanes spiza	Ft,Fh	C	S,M	U	si
Cyanerpes caeruleus	Ft,Fh	C	S,G,M	C	si
*C. cyaneus**	Ft	C	M	R	si
Tersina viridis	Z,Rm	C	G	U	t
PARULIDAE (6)					
*Dendroica striata**	C,Z	C	M	R(Mn)	si
Geothlypis aequinoctialis	M,C	U	S	F	sp,t
Oporornis agilis	C	U	S	R(Mn)	sp
*Wilsonia canadensis**	Ft	U	M	R(Mn)	si
Basileuterus rivularis	Ft,Fh,Fsm	T,U	S	F	t
*Coereba flaveola**	C	Sc	M	R	si
ICTERIDAE (14)					
Molothrus bonariensis	Rm	T	G	R	si
Scaphidura oryzivora	Ft,Rm,S	T,C	S,G	F	t
Clypicterus oseryi	Ft	C	S,G	R	t
Psarocolius decumanus	Ft,Fh	C	G,M	F	t
P. angustifrons	Ft,Z	Sc,C	G,M	C	t
Gymnostinops yuracares	Ft,Fs	C	G,M	F	t

Habitats

Fh	Upland Forest
Ft	Floodplain forest
Fs	Swamp forest
Fsm	Forest stream margins
Fo	Forest openings
Fe	Forest edges
Z	"Zabolo"
B	Bamboo
W	"Willow Bar"
C	Clearing
R	River
Rm	River margins
S	Shores
L	Lake
Lm	Lake margins
M	Marshes
A	Aguajales
O	Overhead

Foraging Position

T	Terrestrial
U	Undergrowth
Sc	Subcanopy
C	Canopy
W	Water
A	Aerial

Sociality

S	Solitary or in pairs
G	Gregarious
M	Mixed-species flocks
A	Army ant followers

Abundance

C	Common
F	Fairly common
U	Uncommon
R	Rare
(M)	Migrant
(Mn)	Migrant from north
(Ms)	Migrant from south
(Mm)	Migrant from Andes
(W)	Wet season only

Evidence

sp	Specimen
t	Tape
si	Species ID by sight
ph	Photo
*	recorded ≤ 3 times

	Habitats	Foraging	Sociality	Abundance	Evidence
Cacicus cela	Ft,Z	Sc,C	G,M	C	t
C. haemorrhous	Ft,Fh	Sc,C	S,G,M	U	t
C. solitarius	Z	U,Sc	S	F	t
Agelaius xanthophthalmus	M	U	S	R	t
Icterus cayanensis	Ft,C	C	S,M	F	t
I. icterus	Lm,M,C	U,Sc	S	U	ph,t
*Leistes superciliaris**	C	T,U	S,G	R(Ms)	si
Dolichonyx oryzivorus	C	U	S,G	R(Mn)	si
Total:	**572 species**				
Resident Forest Species:	**ca. 322 species**				

* recorded 3 or fewer times, and only by sight (further verification required)

Birds of the Lower Río Heath, Including the Pampas del Heath, Bolivia/Perú

Theodore A. Parker, III, Thomas S. Schulenberg and Walter Wust

	Habitats	Foraging	Sociality	Abundance	Evidence
TINAMIDAE (8)					
Tinamus major	Ft	T	S	F	t*
T. guttatus	Fh,Ft	T	S	F	sp,t
Crypturellus cinereus	Ft	T	S	C	t*
C. soui	Ft	T	S	F	sp,t*
C. undulatus	Z,Ft	T	S	C	t*
C. bartletti	Ft	T	S	U	sp,t*
C. parvirostris	Gf	T	S	U	t*
Rhynchotus rufescens	P	T	S	U?	sp
PHALACROCORACIDAE (1)					
Phalacrocorax brasiliensis	R	W	S,G	U	si*
ANHINGIDAE (1)					
Anhinga anhinga	R	W	S	U	si*
ARDEIDAE (7)					
Pilherodias pileatus	S,M	W	S	F	si*
Ardea cocoi	S	W	S	U	si*
Casmerodius albus	S	W	S	U	si*
Bubulcus ibis	C,S	T	S,G	F	si*
Egretta thula	S	W	S	F	si*
Butorides striatus	M	W	S	R	si*
Tigrisoma lineatum	M,Fsm	W	S	U	t*
CICONIIDAE (2)					
Mycteria americana	S,M	W	G,S	U	si*
Jabiru mycteria	S,M,P	W	S	R	si*
THRESKIORNITHIDAE (2)					
Mesembrinibis cayennensis	Fsm	T,W	S	U	si*
Ajaia ajaja	S	W	G	R	si*
ANHIMIDAE (1)					
Anhima cornuta	M,S	T	S	F	t*
ANATIDAE (2)					
Neochen jubata	S	T	S	U	si*

Habitats

Fh	Upland Forest
Ft	Floodplain forest
Gf	Gallery forest
Fs	Swamp forest
Fsm	Forest stream margins
Z	"Zabolo"
B	Bamboo
R	River
Rm	River margins
S	Shores
Lm	Lake Margins
M	Marshes
A	Aguajales
P	Pantanal-like grassland
O	Overhead

Foraging Position

T	Terrestrial
U	Undergrowth
Sc	Subcanopy
C	Canopy
W	Water
A	Aerial

Sociality

S	Solitary or in pairs
G	Gregarious
M	Mixed-species flocks
A	Army ant followers

Abundance

C	Common
F	Fairly common
U	Uncommon
R	Rare
(M)	Migrant, origin unknown
(Mn)	Migrant from north
(Ms)	Migrant from south

Evidence

sp	Specimen
t	Tape
si	Species ID by sight
ph	Photo
*	species observed on both the Bolivian and Peruvian sides of the Río Heath; those without * were noted in Perú only

	Habitats	Foraging	Sociality	Abundance	Evidence
Cairina moschata	R,M	W	S	F	si*
CATHARTIDAE (5)					
Sarcoramphus papa	Ft	T	S	F	si*
Coragyps atratus	S,Z,Ft	T	G	U	si*
Cathartes aura	S,Z,Gf	T	S	R	si
C. burrovianus	P,M	T	S	F	ph
C. melambrotus	Z,Ft,Fh	T	S	F	si
ACCIPITRIDAE (19)					
Gampsonyx swainsonii	Rm,Z	A	S	R	si
Elanoides forficatus	Ft	A	S,G	R	si*
Leptodon cayanensis	Ft,Fh	C,Sc	S	U	si*
Chondrohierax uncinatus	Ft	C	S	R	si*
Harpagus bidentatus	Fh	Sc	S	U	t*
Ictinia plumbea	Ft,Fh	A	S,G	F	si*
Accipiter bicolor	Ft	Sc,C	S	R	si
Leucopternis kuhli	Fh	Sc,C	S	R	si
L. schistacea	Ft	T,Sc	S	U	t*
Asturina nitida	Ft,Gf	C,T	S	U	si
Busarellus nigricollis	Fs,A	T,W	S	U	t*
Buteogallus urubitinga	Rm,S,Z	T,W	S	F	si*
Buteo magnirostris	Rm,Z,Gf	T,C	S	F	t*
B. albicaudatus	P	T	S	R	si
Morphnus guianensis	Ft,Fh	C,Sc	S	R	si
Harpyia harpyja	Ft,Fh	C,Sc	S	R	sp,si
Spizaetus tyrannus	Z,Ft	C,T	S	U	t*
Spizaetus ornatus	Ft	C	S	U	t*
Geranospiza caerulescens	Ft	T,C	S	U	si*
PANDIONIDAE (1)					
Pandion haliaetus	R	W	S	R(Mn)	si*
FALCONIDAE (10)					
Herpetotheres cachinnans	Ft,Z	C,T	S	F	t
Micrastur semitorquatus	Ft?	U,C	S	R	sp
M. ruficollis	Fh	U,C	S	R	si*
M. gilvicollis	Ft,Fh	U,C	S	U	t
Daptrius ater	Rm,S	T,C	S,G	U	t*

	Habitats	Foraging	Sociality	Abundance	Evidence
D. americanus	Ft,Fh	Sc,C	G	F	t*
Milvago chimachima	P	T	S	U	sp,t
Polyborus plancus	P	T	S	R	t
Falco rufigularis	Rm,Ft	A	S	F	t*
F. femoralis	P	A	S	R	si
CRACIDAE (4)					
Ortalis motmot	Z,Ft,Gf	Sc,C	G	C	sp,t*
Penelope jacquacu	Ft,Fh,Gf	C,T	S,G	F	sp,t*
Pipile pipile	Ft,Z,Gf	C,T	S	C	sp,si*
Mitu tuberosa	Ft,Fsm,Gf	T,Sc	S	F	sp,t
PHASIANIDAE (1)					
Odontophorus stellatus	Ft,Fh	T	G	U	sp,t*
RALLIDAE (4)					
Aramides cajanea	Ft,Fsm	T	S	F	si
Porzana albicollis	P	T,U	S	C	sp
Laterallus exilis	M,P	T,U	S	F	t
Micropygia schomburgkii	P	T	S	F	sp
HELIORNITHIDAE (1)					
Heliornis fulica	R	W	S	U	si*
EURYPYGIDAE (1)					
Eurypyga helias	Ft,Fsm,S	T,W	S	U	t*
ARAMIDAE (1)					
Aramus guarauna	M	T,W	S	R	si
PSOPHIIDAE (1)					
Psophia leucoptera	Fh,Ft	T	G	F	si
CHARADRIIDAE (2)					
Hoploxypterus cayanus	S	T	S	F	t*
Charadrius collaris	S	T	S	U	t*
JACANIDAE (1)					
Jacana jacana	M	T,W	S,G	U	si
SCOLOPACIDAE (5)					
Tringa melanoleuca	S	W	S,G	F(Mn)	t*
T. flavipes	S	T,W	S,G	F(Mn)	t*
T. solitaria	S	T,W	S	C(Mn)	t*
Actitis macularia	S	T	S	C(Mn)	t*

Habitats

Fh	Upland Forest
Ft	Floodplain forest
Gf	Gallery forest
Fs	Swamp forest
Fsm	Forest stream margins
Z	"Zabolo"
B	Bamboo
R	River
Rm	River margins
S	Shores
Lm	Lake Margins
M	Marshes
A	Aguajales
P	Pantanal-like grassland
O	Overhead

Foraging Position

T	Terrestrial
U	Undergrowth
Sc	Subcanopy
C	Canopy
W	Water
A	Aerial

Sociality

S	Solitary or in pairs
G	Gregarious
M	Mixed-species flocks
A	Army ant followers

Abundance

C	Common
F	Fairly common
U	Uncommon
R	Rare
(M)	Migrant, origin unknown
(Mn)	Migrant from north
(Ms)	Migrant from south

Evidence

sp	Specimen
t	Tape
si	Species ID by sight
ph	Photo
*	species observed on both the Bolivian and Peruvian sides of the Rio Heath; those without * were noted in Perú only

	Habitats	Foraging	Sociality	Abundance	Evidence
Calidris melanotos	S	T	G	F(Mn)	t*
LARIDAE (2)					
Phaetusa simplex	R	W,A	S	U	si*
Sterna superciliaris	R	W,A	S	F	si*
RHYNCHOPIDAE (1)					
Rynchops niger	R	W	S	R	si*
COLUMBIDAE (9)					
Columba speciosa	Gf	C	S,G	F	t
C. cayennensis	Gf,Z	U,C	S,G	C	sp,t*
C. subvinacea	Ft	C	S	U	t*
C. plumb a	Ft,Fh	C	S	C	t*
Columbina talpacoti	Z,Rm	T	S,G	F	t*
C. picui	Z,Rm	T	S,G	U(Ms?)	si*
Claravis pretiosa	Z,Ft	T	S	R	si*
Leptotila rufaxilla	Z	T	S	F	t*
Geotrygon montana	Fh,Ft	T	S	F	sp,t
PSITTACIDAE (20)					
Ara ararauna	Ft,A	C	G	C	sp,t*
A. macao	Ft,Fh	C	S,G	F	t*
A. chloroptera	Ft,Fh	C	S,G	F	sp,t*
A. severa	Z,Ft,A	C	S,G	C	sp,t*
A. manilata	A	C	G	C	sp,t*
A. couloni	Ft	C	G	R	si*
A. nobilis	Gf,A	C	G	C	sp
Aratinga leucophthalmus	Z,Ft,Gf	C	G	U	t*
A. weddellii	Z,Ft	C	G	C	t*
A. aurea	Gf,P	C	S,G	F	sp*
Pyrrhura rupicola	Ft,Fh	C	G	U	t*
Forpus sclateri	Ft,Fh	C	S	U	t*
Brotogeris cyanoptera	Z,Ft,Fh	C	G	C	t*
Nannopsittaca dachilleae	Ft	C	S	R	t*
Touit huetii	Ft?	C	G	R	t*
Pionites leucogaster	Ft,Fh	C	G	C	sp,t*
Pionopsitta barrabandi	Fh	Sc,C	S,G	F	t*
Pionus menstruus	Ft,Z,Gf	C	S,G	F	t*

	Habitats	Foraging	Sociality	Abundance	Evidence
Amazona ochrocephala	Gf,Z,Ft	C	S	C	t*
A. farinosa	Ft,Fh	C	S,G	C	sp,t*
CUCULIDAE (7)					
Coccyzus melacoryphus	Z,Gf	C	S,M	U(Ms)	sp
Piaya cayana	Z,Ft,Fh,Gf	Sc,C	S,M	F	t
P. minuta	Z	U	S	R	t
Crotophaga ani	P,Rm	U,T	G	F	t
Tapera naevia	P	U	S	R	sp
Dromococcyx phasianellus	Ft	T,U	S	R	t*
D. pavoninus	Ft	U	S	R	t*
OPISTHOCOMIDAE (1)					
Opisthocomus hoazin	Rm,M	U,C	G	U	si
STRIGIDAE (8)					
Otus choliba	Gf	Sc,C	S	U	sp,t
O. watsonii	Ft,Fh	U,C	S	C	t*
Lophostrix cristata	Ft,Fh	Sc,C	S	U	t*
Pulsatrix perspicillata	Ft	Sc,C	S	U	t*
Glaucidium hardyi	Ft,Fh	C,Sc	S	F	t*
G. brasilianum	Gf,Z	C,Sc	S	U	t
Ciccaba (virgata)	Ft,Fh	C,Sc	S	U	si
Rhinoptynx clamator	P	Sc,C?	S	R	sp
NYCTIBIIDAE (2)					
Nyctibius grandis	Ft	A,C	S	U	t
N. griseus	Gf,Ft	A,C	S	F	sp,t
CAPRIMULGIDAE (10)					
Chordeiles rupestris	S	A	G	C	t*
Podager nacunda	P	A	G	R(Ms?)	si
Nyctidromus albicollis	Z	A	S	F	sp,t*
Nyctiphrynus ocellatus	Ft	A,Sc	S	U	sp,t*
Caprimulgus rufus	Gf	A	S	R(Ms)	si
C. sericocaudatus	Gf,Ft	A,C	S	R(Ms?)	t
C. maculicaudus	P	A	S	C	sp
C. parvulus	Gf	A	S	U(Ms)	t
Hydropsalis climacocerca	Rm,S	A	S	F	t*
H. brasiliana	P	A	S	F(Ms)	sp

Habitats

Fh	Upland Forest
Ft	Floodplain forest
Gf	Gallery forest
Fs	Swamp forest
Fsm	Forest stream margins
Z	"Zabolo"
B	Bamboo
R	River
Rm	River margins
S	Shores
Lm	Lake Margins
M	Marshes
A	Aguajales
P	Pantanal-like grassland
O	Overhead

Foraging Position

T	Terrestrial
U	Undergrowth
Sc	Subcanopy
C	Canopy
W	Water
A	Aerial

Sociality

S	Solitary or in pairs
G	Gregarious
M	Mixed-species flocks
A	Army ant followers

Abundance

C	Common
F	Fairly common
U	Uncommon
R	Rare
(M)	Migrant, origin unknown
(Mn)	Migrant from north
(Ms)	Migrant from south

Evidence

sp	Specimen
t	Tape
si	Species ID by sight
ph	Photo
*	species observed on both the Bolivian and Peruvian sides of the Río Heath; those without * were noted in Perú only

	Habitats	Foraging	Sociality	Abundance	Evidence
APODIDAE (6)					
Streptoprocne zonaris	O	A	G	U	sp*
Chaetura cinereiventris	O(Ft)	A	G,M	F(M?)	t*
C. egregia	O(Ft)	A	G,M	U(Ms?)	t*
C. brachyura	O(Gf)	A	G,M	R	si*
Panyptila cayennensis	O(Ft)	A	S,M	U	t*
Tachornis squamata	A,P,Ft	A	S,G	C	t*
TROCHILIDAE (14)					
Glaucis hirsuta	Z,Ft	U,Sc	S	U	sp,t
Threnetes leucurus	Ft	U	S	?	sp
Phaethornis philippi	Fh	U	S	F	sp
P. hispidus	Z,Ft	U	S	F	sp,t
P. ruber	Ft,Fh	U	S	C	t
Eupetomena macroura	Gf,P	A,C	S	F	sp
Florisuga mellivora	Ft,Fh	Sc,C	S	U	sp
Anthracothorax nigricollis	Rm,Gf	C	S	U	sp
Thalurania furcata	Ft,Fh	U,Sc	S	F	sp*
Hylocharis cyanus	Ft,Gf	U,C	S	F	t*
Polytmus guainumbi	P	U,Sc	S	U	sp
P. theresiae	P,Gf	U,Sc	S	U	sp
Amazilia lactea	Gf,Z	C	S	R	si
Heliomaster longirostris	Z,Gf	C	S	R	si*
TROGONIDAE (6)					
Pharomachrus pavoninus	Ft,Fh	C	S	U	sp,t*
Trogon melanurus	Ft,Fh,Gf	C,Sc	S	F	sp,t*
T. viridis	Ft,Fh,Gf	C,Sc	S,M	C	t*
T. collaris	Ft	U,Sc	S,M	F	sp,t*
T. curucui	Z,Ft	C,Sc	S,M	U	t*
T. violaceus	Ft,Fh	C,Sc	S,M	U	t
ALCEDINIDAE (5)					
Ceryle torquata	Rm	W	S	F	t*
Chloroceryle amazona	Rm	W	S	F	t*
C. americana	Fsm,Rm	W	S	U	si*
C. inda	Fsm	W	S	U	sp
C. aenea	Gf,Fsm	W	S	U	sp

	Habitats	Foraging	Sociality	Abundance	Evidence
MOMOTIDAE (3)					
Electron platyrhynchum	Ft,Fh	C,Sc	S	C	t*
Baryphthengus martii	Ft	C,Sc	S	F	t*
Momotus momota	Ft,Fh	Sc	S	U	sp,t
GALBULIDAE (4)					
Brachygalba albogularis	Z	A,C	S	R	si*
Galbula cyanescens	Z,Gf,Ft	Sc,U	S,M	F	sp,t*
G. dea	Ft,Fh	A,C	S,M	U	sp
Jacamerops aurea	Ft,Fh	Sc,C	S	U	sp,t*
BUCCONIDAE (8)					
Notharchus macrorhynchos	Ft,Fh	C	S	U	t
Bucco macrodactylus	Z,Ft	Sc,C	S	U	si
Nystalus chacuru	Gf,P	C,T	S	F	sp
N. striolatus	Ft	C,Sc	S	U	t*
Malacoptila semicincta	Fh	U	S	U	sp
Monasa nigrifrons	Z,Ft	C,Sc	G,M	C	sp,t*
M. morphoeus	Fh	C,Sc	G,M	F	sp,t
Chelidoptera tenebrosa	Rm,Z	A,C	S	C	t*
CAPITONIDAE (2)					
Capito niger	Ft,Fh	C,Sc	S,M	F	sp,t*
Eubucco richardsoni	Ft	C	S,M	U	t*
RAMPHASTIDAE (8)					
Aulacorhynchus prasinus	Ft	C	S	U	t*
Pteroglossus castanotis	Ft,Z	C	G	F	t*
P. inscriptus	Ft,Z	C,Sc	G	R	t
P. mariae	Ft,Fh	C	G	F	sp,t*
Pteroglossus beauharnaesii	Ft,Fh	C	G	U	sp,t*
Ramphastos culminatus	Ft,Fh	C	S,G	C	sp,t*
R. cuvieri	Ft,Fh	C	S,G	C	sp,t*
R. toco	Gf	C	S	R	t
PICIDAE (15)					
Picumnus rufiventris	Z	U,Sc	S,M	R	sp
P. aurifrons	Ft	C,Sc	S,M	U	si
Chrysoptilus punctigula	Z	C,Sc	S	R	t*
Piculus chrysochloros	Ft	C	S,M	U	t?*

Habitats

Fh	Upland Forest
Ft	Floodplain forest
Gf	Gallery forest
Fs	Swamp forest
Fsm	Forest stream margins
Z	"Zabolo"
B	Bamboo
R	River
Rm	River margins
S	Shores
Lm	Lake Margins
M	Marshes
A	Aguajales
P	Pantanal-like grassland
O	Overhead

Foraging Position

T	Terrestrial
U	Undergrowth
Sc	Subcanopy
C	Canopy
W	Water
A	Aerial

Sociality

S	Solitary or in pairs
G	Gregarious
M	Mixed-species flocks
A	Army ant followers

Abundance

C	Common
F	Fairly common
U	Uncommon
R	Rare
(M)	Migrant, origin unknown
(Mn)	Migrant from north
(Ms)	Migrant from south

Evidence

sp	Specimen
t	Tape
si	Species ID by sight
ph	Photo
*	species observed on both the Bolivian and Peruvian sides of the Río Heath; those without * were noted in Perú only

	Habitats	Foraging	Sociality	Abundance	Evidence
Celeus elegans	Ft	Sc,C	S	U	sp
C. grammicus	Fh	C	S,M	F	sp,t*
C. flavus	Ft	C,Sc	S,G	U	t*
C. torquatus	Ft	C,Sc	S	R	t*
Dryocopus lineatus	Z,Ft,Gf	Sc,C	S	F	sp,t*
Melanerpes cruentatus	Ft,Fh	C	S,M	C	t
Leuconerpes candidus	Gf,P	T,C	S	R	si
Veniliornis passerinus	Z	Sc,C	S,M	F	sp,t*
V. affinis	Ft,Fh	C,Sc	M	F	sp,t*
Campephilus melanoleucus	Ft,Z	C,Sc	S	F	t*
C. rubricollis	Fh	U,C	S	F	sp,t
DENDROCOLAPTIDAE (14)					
Dendrocincla fuliginosa	Ft	U,Sc	S,M,A	F	sp,t
D. merula	Fh	U,Sc	S,M,A	U	sp
Deconychura longicauda	Ft,Fh	C,Sc	S,M	U	sp,t*
Sittasomus griseicapillus	Ft	U,Sc	M	F	sp,t*
Glyphorhynchus spirurus	Ft,Fh	U,C	S,M	U	sp
Dendrexetastes rufigula	Ft	C	S,M	F	t*
Dendrocolaptes certhia	Ft,Fh	Sc,C	S,M,A	F	sp,t*
D. picumnus	Ft,Fh	U,C	S,A	U	t*
Xiphorhynchus picus	Z,Gf	Sc	S,M	U	t*
X. obsoletus	Ft	Sc	S,M	R	si*
X. spixii	Ft,Fh	U,Sc	M	C	sp
X. guttatus	Ft,Fh	U,C	S,M	C	sp
Lepidocolaptes albolineatus	Ft,Fh	C	M	U	t
Campylorhamphus trochilirostris	Ft	U	S,M	R	si
FURNARIIDAE (20)					
Furnarius leucopus	Z,Ft	T	S	U	t*
Synallaxis hypospodia	P	U	S	C	sp
S. albescens	P,Rm	U	S	U(Ms?)	t
S. gujanensis	Z	U	S	F	sp,t
S. rutilans	Fh	T,U	S	F	sp
Cranioleuca gutturata	Ft	C,Sc	M	U	sp
Thripophaga fusciceps	Ft	Sc	S,M	U	t*
Berlepschia rikeri	A	C	S	F	t

	Habitats	Foraging	Sociality	Abundance	Evidence
Ancistrops strigilatus	Ft,Fh	C,Sc	M	F	sp,t*
Philydor erythrocercus	Fh,Ft	Sc,C	M	U	sp
P. pyrrhodes	Ft	Sc,U	S,M	U	t
P. rufus	Z	C	S,M	U	t*
P. erythropterus	Ft,Fh	C	M	F	t*
P. ruficaudatus	Ft	C,Sc	M	U	t*
Automolus infuscatus	Fh	U	M	F	sp
A. ochrolaemus	Ft	U	S,M	F	sp,t
A. rufipileatus	Z,Ft	U	S,M	F	sp,t*
Xenops milleri	Ft	C	M	R	si
X. minutus	Ft,Fh	U,Sc	M	F	sp,t
Sclerurus caudacutus	Ft,Fh	T	S	U	t
FORMICARIIDAE (44)					
Cymbilaimus lineatus	Ft,Fh	C,Sc	S,M	F	t
C. sanctaemariae	Ft,B	Sc	S,M	F	t*
Frederickena unduligera	Ft	U	S	R	sp
Taraba major	Z	U	S	U	t*
Thamnophilus doliatus	P,Z	U,Sc	S,M	F	sp,t*
T. aethiops	Ft,Fh	U	S	F	sp
T. schistaceus	Ft,Fh	Sc	M	C	sp*
Pygiptila stellaris	Ft,Fh	C	M	F	sp,t
Thamnomanes ardesiacus	Ft,Fh	U	M	U	sp
T. schistogynus	Ft,Z	U,Sc	M	F	t*
T. amazonicus	Gf	U,Sc	S,M	F	t
Myrmotherula brachyura	Ft, Fh	C	M	C	t*
M. sclateri	Ft,Fh	C	M	C	t*
M. surinamensis	Ft	Sc	S,M	U	t*
M. hauxwelli	Ft,Fh	U	S,M	F	sp
M. ornata	Ft,B	Sc	M	U	t*
M. leucophthalma	Ft,Fh	U	M	U	t?
M. axillaris	Ft,Fh	U,Sc	M	C	sp*
M. longipennis	Fh	U,Sc	M	R?	sp
M. menetriesii	Ft,Fh	Sc,C	M	C	sp*
Dichrozona cincta	Fh	T	S	R?	t
Formicivora rufa	P	U	S	F	sp

Habitats

Fh	Upland Forest
Ft	Floodplain forest
Gf	Gallery forest
Fs	Swamp forest
Fsm	Forest stream margins
Z	"Zabolo"
B	Bamboo
R	River
Rm	River margins
S	Shores
Lm	Lake Margins
M	Marshes
A	Aguajales
P	Pantanal-like grassland
O	Overhead

Foraging Position

T	Terrestrial
U	Undergrowth
Sc	Subcanopy
C	Canopy
W	Water
A	Aerial

Sociality

S	Solitary or in pairs
G	Gregarious
M	Mixed-species flocks
A	Army ant followers

Abundance

C	Common
F	Fairly common
U	Uncommon
R	Rare
(M)	Migrant, origin unknown
(Mn)	Migrant from north
(Ms)	Migrant from south

Evidence

sp	Specimen
t	Tape
si	Species ID by sight
ph	Photo
*	species observed on both the Bolivian and Peruvian sides of the Río Heath; those without * were noted in Perú only

	Habitats	Foraging	Sociality	Abundance	Evidence
Terenura humeralis	Ft,Fh	C	M	F	t?
Drymophila devillei	Ft,B	Sc	S,M	U	t
Cercomacra cinerascens	Ft,Fh	Sc	S,M	F	t*
C. nigrescens	Z,B	U	S	U	t
C. manu	Ft,B	Sc	S	U	t
Myrmoborus leucophrys	Ft	U	S	F	sp,t*
M. myotherinus	Ft,Fh	U	S,A	C	sp,t
Hypocnemis cantator	Ft	U	S,M	F	sp,t
Hypocnemoides maculicauda	Fsm	T,U	S	U	t
Percnostola lophotes	Z,B	T,U	S	F	t*
Myrmeciza hemimelaena	Ft,Fh	T,U	S	C	sp*
M. hyperythra	Ft	T,U	S	U	sp,t*
M. goeldii	Ft,B	T,U	S,A	R	t
M. atrothorax	Z	T,U	S	U	sp,t*
Gymnopithys salvini	Ft,Fh	U	A	F	sp
Hylophylax poecilinota	Ft,Fh	U	S,A	U	sp,t
Phlegopsis nigromaculata	Ft	T,U	S,A	U	t*
Formicarius colma	Fh	T	S	F	sp,t
F. analis	Ft	T	S,A	C	sp,t*
Hylopezus berlepschi	Z	T	S	U	t
Myrmothera campanisona	Ft,Fh	T	S	U	t
Conopophaga peruviana	Ft	U,T	S	R	si
COTINGIDAE (5)					
Iodopleura isabellae	Ft	C	S	R	sp
Lipaugus vociferans	Ft,Fh	Sc,C	S	C	sp,t*
Cotinga maynana	Z,Ft	C	S	F	si*
Gymnoderus foetidus	Ft	C	S,G	F	si*
Querula purpurata	Fh,Ft	C	G	U	t
PIPRIDAE (12)					
Schiffornis major	Ft	U	S	U	t*
S. turdinus	Fh	U	S	U	sp
Piprites chloris	Ft,Fh	Sc,C	M	F	t*
Xenopipo atronitens	Gf	U,Sc	S	R?	sp
Heterocercus linteatus	Ft	Sc	S	R	si
Tyranneutes stolzmanni	Ft,Fh	Sc,C	S	C	t*

	Habitats	Foraging	Sociality	Abundance	Evidence
Manacus manacus	Ft,Gf	U	S	U	sp
Machaeropterus pyrocephalus	Ft	Sc,C	S	F	sp
Pipra coronata	Fh	U	S	R	si
P. fasciicauda	Ft	Sc,U	S	F	sp,t
P. rubrocapilla	Fh	Sc,U	S	F	sp
P. chloromeros	Ft	Sc,U	S	U	sp
TYRANNIDAE (67)					
Zimmerius gracilipes	Ft,Fh	C	S,M	F	t*
Ornithion inerme	Ft,Fh	C	S,M	F	t*
Camptostoma obsoletum	Z,Gf	C	S,M	U	sp,t
Sublegatus obscurior	Z	Sc,C	S,M	R	si*
Phaeomyias murina	Z	C	S,M	U	t*
Tyrannulus elatus	Ft,Fh	C	S,M	F	t*
Myiopagis gaimardii	Ft,Fh	C	M	C	sp,t*
M. caniceps	Ft	C	M	U	sp,t*
M. viridicata	Ft,Fh,Z	Sc,C	M	U(Ms?)	t*
Elaenia flavogaster	P,Gf	C	S	C	sp
E. spectabilis	Z	C,Sc	S,M	U(Ms)	si*
E. parvirostris	Z	C	S,M	U(Ms)	si*
E. cristata	P	C	S	U	sp
Inezia inornata	Z,Gf	Sc,C	G,M	F(Ms?)	si*
Euscarthmus meloryphus	Gf,Z	U	S	R(Ms?)	t
Mionectes oleagineus	Ft,Fh	Sc,U	M	U	sp
M. macconnelli	Ft,Fh	Sc,U	M	R	sp
Leptopogon amaurocephalus	Ft	U,Sc	M,S	F	sp,t*
Corythopis torquata	Ft,Fh	T	S	F	sp,t
Myiornis ecaudatus	Ft	C	S	F	t
Hemitriccus zosterops	Ft,Fh	Sc	S	F	sp,t
H. iohannis	Ft	Sc	S	U	t*
H. striaticollis	Gf	Sc,C	S	F	sp
Todirostrum latirostre	Z	U	S	F	t*
T. maculatum	Z	C,Sc	S,M	F	t*
T. chrysocrotaphum	Ft,Fh	C	S,M	F	t*
Ramphotrigon ruficauda	Ft,Fh	Sc	S	F	sp,t*
Tolmomyias sulphurescens	Ft	C	M	U	si*

Habitats

Fh	Upland Forest
Ft	Floodplain forest
Gf	Gallery forest
Fs	Swamp forest
Fsm	Forest stream margins
Z	"Zabolo"
B	Bamboo
R	River
Rm	River margins
S	Shores
Lm	Lake Margins
M	Marshes
A	Aguajales
P	Pantanal-like grassland
O	Overhead

Foraging Position

T	Terrestrial
U	Undergrowth
Sc	Subcanopy
C	Canopy
W	Water
A	Aerial

Sociality

S	Solitary or in pairs
G	Gregarious
M	Mixed-species flocks
A	Army ant followers

Abundance

C	Common
F	Fairly common
U	Uncommon
R	Rare
(M)	Migrant, origin unknown
(Mn)	Migrant from north
(Ms)	Migrant from south

Evidence

sp	Specimen
t	Tape
si	Species ID by sight
ph	Photo
*	species observed on both the Bolivian and Peruvian sides of the Río Heath; those without * were noted in Perú only

	Habitats	Foraging	Sociality	Abundance	Evidence
T. assimilis	Ft,Fh	C	M	F	sp,t
T. poliocephalus	Ft,Z	C	S,M	U	t
T. flaviventris	Z	C,Sc	S,M	F	t
Platyrinchus coronatus	Ft	Sc,U	S	F	sp
Onychorhynchus coronatus	Ft,Fh,Fsm	U,Sc	S,M	U	sp
Terenotriccus erythrurus	Ft,Fh	U,C	S,M	U	sp,t*
Myiophobus fasciatus	Gf,Z	U	S,M	F	sp,t
Lathrotriccus euleri	Ft,Z,B	U,Sc	S	F	sp,t
Cnemotriccus fuscatus	Gf,Z	U,Sc	S	U	t*
Pyrocephalus rubinus	Rm,Z,Gf	C,A	S	F(Ms)	sp*
Ochthoeca littoralis	Rm,S	T,A	S	F	sp,t*
Xolmis cinerea	P	A,T	S	F	sp
Muscisaxicola fluviatilis	Rm,S	T	S	U	si*
Fluvicola pica	Rm,S	T	S	R(Ms)	si*
Satrapa icterophrys	Rm,Z	C,Sc	S,M	R(Ms)	si*
Attila bolivianus	Ft	C,Sc	S,M	F	t*
A. spadiceus	Ft,Fh	C,Sc	S,M	F	t*
Rhytipterna simplex	Ft,Fh	Sc,C	S,M	C	sp,t*
Laniocera hypopyrra	Ft,Fh	Sc,C	S,M	U?	sp
Sirystes sibilator	Ft	C	S,M	F	t*
Myiarchus swainsoni	Gf,Z,Ft	C	M	U(Ms)	sp
M. ferox	Z,C	Sc,C	S,M	F	sp,t*
M. tyrannulus	Ft,Fh	C	S,M	F(Ms)	t
Pitangus sulphuratus	Rm,Z	U,C	S	F	sp,t*
Megarhynchus pitangua	C,Z	Sc,C	S	F	t*
Myiozetetes cayanensis	Lm	U,C	S	U	si*
M. similis	Rm,Z	U,C	S,G	C	t*
M. granadensis	Z,Ft	Sc,C	S,G	F	t*
M. luteiventris	Ft	C	S,G	U	t*
Myiodynastes maculatus	Ft	C,Sc	M	U(Ms)	si
Legatus leucophaius	Ft,Gf	A,C	U	U	t*
Empidonomus varius	Ft	C	M	R(Ms)	si*
Tyrannopsis sulphurea	A	C	S	U	sp
Tyrannus albogularis	A	A,C	S	U	t
T. melancholicus	Rm,Z,Gf	A,C	S	F	t*

	Habitats	Foraging	Sociality	Abundance	Evidence
Pachyramphus polychopterus	Z	C	S,M	F	t*
P. marginatus	Ft,Fh	C	M	F	t*
Tityra cayana	Ft	C	S	F	si*
T. inquisitor	Ft	C	S	U	si*
HIRUNDINIDAE (6)					
Phaeoprogne tapera	R	A	S,G	F	t*
Progne chalybea	R	A	S,G	U	si*
Tachycineta albiventer	R	A	S,G	C	t*
Notiochelidon cyanoleuca	R	A	G	R(Ms?)	si
Atticora fasciata	R	A	G	C	t*
Stelgidopteryx ruficollis	R	A	S,G	C	t*
CORVIDAE (1)					
Cyanocorax violaceus	Z,Ft	C,Sc	G,M	C	t*
TROGLODYTIDAE (7)					
Campylorhynchus turdinus	Z	C,Sc	S,M	F	t*
Thryothorus genibarbis	Z,Ft,B	U	S	C	sp,t*
T. leucotis	Z	U	S	U	t*
Troglodytes aedon	Rm,C	U	S	U	t
Microcerculus marginatus	Ft,Fh	T	S	F	t*
Cyphorhinus aradus	Ft,Fh	T	S,M	F	sp,t
Donacobius atricapillus	Lm	U	S	F	t*
TURDIDAE (6)					
Turdus leucomelas	Gf	T,C	S	U(Ms?)	sp
T. amaurochalinus	Z,Gf,Ft	T,C	S	F(Ms?)	sp,t*
T. ignobilis	Z	T,C	S,G	F?	si*
T. lawrencii	Ft	T,C	S	F	t*
T. hauxwelli	Ft	T	S	U	t*
T. albicollis	Ft,Fh	T,Sc	S	F	sp,t
MOTACILLIDAE (1)					
Anthus lutescens	P	T	S	U	si
VIREONIDAE (4)					
Cyclarhis gujanensis	Z	C	S,M	R	sp,t
Vireo olivaceus	Ft,Fh,Z,Gf	C	G,M	C(Ms)	t*
Hylophilus hypoxanthus	Ft,Fh	C	M	C	sp,t*
Hylophilus sp.	Ft	C	M	U?	t

Habitats

Fh	Upland Forest
Ft	Floodplain forest
Gf	Gallery forest
Fs	Swamp forest
Fsm	Forest stream margins
Z	"Zabolo"
B	Bamboo
R	River
Rm	River margins
S	Shores
Lm	Lake Margins
M	Marshes
A	Aguajales
P	Pantanal-like grassland
O	Overhead

Foraging Position

T	Terrestrial
U	Undergrowth
Sc	Subcanopy
C	Canopy
W	Water
A	Aerial

Sociality

S	Solitary or in pairs
G	Gregarious
M	Mixed-species flocks
A	Army ant followers

Abundance

C	Common
F	Fairly common
U	Uncommon
R	Rare
(M)	Migrant, origin unknown
(Mn)	Migrant from north
(Ms)	Migrant from south

Evidence

sp	Specimen
t	Tape
si	Species ID by sight
ph	Photo
*	species observed on both the Bolivian and Peruvian sides of the Río Heath; those without * were noted in Perú only

	Habitats	Foraging	Sociality	Abundance	Evidence
EMBERIZINAE (10)					
Myospiza humeralis	P	T,U	S	U	sp
M. aurifrons	S,C	T,U	S	C	t*
Emberizoides herbicola	P	U	S	F	sp
Volatinia jacarina	P,M	T,U	S,G	U	sp
Sporophila plumbea	P	U	S	U	sp
S. caerulescens	P,M	T,U	G,M	C(Ms)	sp*
S. castaneiventris	P,M	U,C	S,M	U	si*
Oryzoborus angolensis	M	U	S	R	si
Coryphaspiza melanotis	P	U	S	F	sp
Paroaria gularis	Rm,Z	U,C	S,M	F	sp*
CARDINALINAE (3)					
Saltator maximus	Z,Ft	C,Sc	S,M	F	t*
S. coerulescens	Z,C	U,C	S	F	t
Cyanocompsa cyanoides	Ft	U	S	F	sp,t*
THRAUPINAE (28)					
Schistochlamys melanopis	P,Gf	U,C	S,G	C	sp
Thlypopsis sordida	Z	U,C	M	R(Ms?)	si
Hemithraupis flavicollis	Ft,Fh	C	M	F	sp,t
Lanio versicolor	Fh	C,Sc	U	F	si
Tachyphonus cristatus	Fh	C	M	U	sp
T. luctuosus	Z,Ft,Fh	C,Sc	M	F	sp,t*
Habia rubica	Fh,Ft	U	G,M	U	sp,t
Ramphocelus carbo	Z,Ft,C	U,C	G,M	C	sp,t*
Thraupis episcopus	Z	C,Sc	S,M	U	si*
T. palmarum	Ft,Gf	C	S,M	F	sp,t*
Euphonia chlorotica	Gf,Ft	C	S	F	t
E. laniirostris	Z	C,Sc	S,M	U	t
E. chrysopasta	Ft	C	S,M	F	t *
E. minuta	Ft	C	M	U	si
E. rufiventris	Ft	C	S,M	F	t*
Chlorophonia cyanea	Ft,Fh	C	M	R(M?)	t
Tangara mexicana	Ft	C	M,G	C	t*
T. chilensis	Ft,Fh	C,Sc	G,M	C	t*
T. schrankii	Fh	U,C	M	C	sp

CONSERVATION INTERNATIONAL

Rapid Assessment Program

	Habitats	Foraging	Sociality	Abundance	Evidence
T. cayana	Gf	C	S	F	sp
T. velia	Fh	C	M	U	sp,t
Dacnis lineata	Ft,Fh	C	M	C	si*
D. flaviventer	Ft	C	S,M	U	t
D. cayana	Ft,Fh	C	M	F	si
Chlorophanes spiza	Ft,Fh	C	S,M	U	sp
Cyanerpes caeruleus	Ft,Fh,Gf	C	S,M	U	sp
Tersina viridis	Z,Ft	C	S,G	U	t*
PARULIDAE (2)					
Geothlypis aequinoctialis	M,Rm	U	S	U	t*
Coereba flaveola	Gf	C	S	U	sp
ICTERIDAE (9)					
Scaphidura oryzivora	Rm,S,P	T,C	S,G	F	t*
Psarocolius decumanus	Ft,Fh	C	G,M	F	sp,t*
P. angustifrons	Ft,Z	C,Sc	G,M	F	t*
P. yuracares	Ft,Fs	C	G,M	F	sp,t*
Cacicus cela	Ft,Z	C,Sc	G,M	C	t*
C. solitarius	Z	U,Sc	S	F	t*
Gnorimopsar chopi	A,Gf,P	C	G	F	sp
Icterus cayanensis	Ft,C	C	S,M	U	sp,t
Icterus icterus	Z?	U,Sc	S	R?	sp

Habitats

Fh	Upland Forest
Ft	Floodplain forest
Gf	Gallery forest
Fs	Swamp forest
Fsm	Forest stream margins
Z	"Zabolo"
B	Bamboo
R	River
Rm	River margins
S	Shores
Lm	Lake Margins
M	Marshes
A	Aguajales
P	Pantanal-like grassland
O	Overhead

Foraging Position

T	Terrestrial
U	Undergrowth
Sc	Subcanopy
C	Canopy
W	Water
A	Aerial

Sociality

S	Solitary or in pairs
G	Gregarious
M	Mixed-species flocks
A	Army ant followers

Abundance

C	Common
F	Fairly common
U	Uncommon
R	Rare
(M)	Migrant, origin unknown
(Mn)	Migrant from north
(Ms)	Migrant from south

Evidence

sp	Specimen
t	Tape
si	Species ID by sight
ph	Photo
*	species observed on both the Bolivian and Peruvian sides of the Río Heath; those without * were noted in Perú only

Mammals of the Upper Tambopata/Távara

Louise H. Emmons and Mónica Romo

A. Mammals collected or identified by expedition members (X), and (W) species seen by Walter Wust prior to this expedition (not including records duplicated on this expedition). Elevation = highest elevation (in m) at which the species was recorded.

	Ccolpa de Guacamayos	Mirador Boca Távara	Cerros del Távara	Elevation
Didelphidae (opossums)				
Didelphis marsupialis	X			
Marmosops noctivagus	X			
Metachirus nudicaudatus	X			
Micoureus cinereus			X	500
Philander opossum	W			
Myrmecophagidae (anteaters)				
Cyclopes didactylus	W			
Myrmecophaga tridactyla	W			
Tamandua tetradactyla	W			
Bradypodidae (three-toed sloths)				
Bradypus variegatus	W			
Dasypodidae (armadillos)				
Dasypus kappleri	X			
Emballonuridae (sheath-tailed bats)				
Saccopteryx bilineata	X*			
Phyllostomidae (leaf-nosed bats)				
Anoura caudifer			X*	700
Artibeus anderseni	X*			
Artibeus jamaicencis	X*			
Artibeus obscurus	X*		X*	
Carollia brevicauda	X*		X	700
Carollia castanea	X*		X	700
Carollia perspicillata	X*		X	700
Chiroderma villosum	X*			
Choeroniscus intermedius	X*			
Desmodus rotundus			X*	
Diphylla ecaudata	X*			
Glossophaga soricina	X*			
Lonchophylla thomasi	X*		X*	700

	Ccolpa de Guacamayos	Mirador Boca Távara	Cerros del Távara	Elevation
Mesophylla macconnelli	X*		X*	700
Micronycteris minuta	X*			
Phyllostomus elongatus	X*			
Phyllostomus hastatus	X*		X	700
Platyrrhinus helleri	X*		X	700
Rhinophylla pumilio	X*			
Sturnira lilium	X		X*	
Sturnira magna			X*	700
Sturnira tildae	X*			
Tonatia brasiliense			X*	
Tonatia bidens	X*		X	700
Trachops cirrhosus	X*			
Uroderma magnirostrum	X*			
Vampyrodes caraccioli			X*	700
Vampyressa cf. *melissa*	X?			
Vespertilionidae (vespertilionid bats)				
Myotis riparius	X*		X*	
Callithrichidae (tamarins)				
Saguinus fuscicollis	X	X		
Cebidae (monkeys)				
Alouatta seniculus	X			
Aotus trivirgatus	X	X	X	550
Ateles paniscus	X	W	X	870
Callicebus moloch	X			
Cebus apella	X		X	550
Cebus albifrons	W			
Lagothrix lagothricha		W		
Saimiri sciureus boliviensis	X			
Procyonidae (raccoon family)				
Potos flavus	X			
Nasua nasua	X			
Mustelidae (weasel family)				
Eira barbara	X			
Lontra longicaudis	X		X	300
Pteronura brasiliensis	X			

*	Specimen collected

	Ccolpa de Guacamayos	Mirador Boca Távara	Cerros del Távara	Elevation
Felidae (cats)[1]				
Puma concolor			X	500
Leopardus pardalis	X			
Leopardus wiedii	X			
Panthera onca	X		X	300
Tapiridae (tapirs)				
Tapirus terrestris	X		X	750
Tayassuidae (peccaries)				
Tayassu pecari	X			
Tayassu tajacu	X		X	
Cervidae (deer)				
Mazama americana	X		X	870
RODENTS				
Sciuridae (squirrels)				
Microsciurus sp.			X	550
Sciurus spadiceus	X		X	
Sciurus ignitus	X		X?	
Muridae (rats and mice)				
Neacomys spinosus			X*	
Oecomys bicolor	X*			
Oecomys sp.	X*			
Oligoryzomys microtis	X*			
Oryzomys capito	X*			
Rhipidomys cf. *couesi*	X*			
Erethizontidae (porcupines)				
Coendou cf. *bicolor*	W			
Hydrochaeridae (capybara)				
Hydrochaeris hydrochaeris	X			
Agoutidae (pacas)				
Agouti paca	X		X	720
Dasyproctidae (agoutis)				
Dasyprocta variegata	X		X	700
Myoprocta pratti	X			

	Ccolpa de Guacamayos	Mirador Boca Távara	Cerros del Távara	Elevation
Echimyidae (spiny rats)				
Dactylomys dactylinus	X			
Mesomys hispidus	X		X	580
Proechimys simonsi	X*		X*	750
Leporidae (rabbits)				
Silvilagus brasiliensis	X			

*   Specimen collected

B. Other mammal species reported to occur along the Río Tambopata by indigenous Ese'eja informants (Sixto and Augustín). Only species not recorded on the expedition are included.

Megalonychidae (two-toed sloths)

Choloepus cf. *hoffmanni*

Dasypodidae (armadillos)

Cabassous unicinctus

Dasypus novemcinctus

Priodontes maximus

Canidae (dogs)

Atelocynus microtis

Speothos venaticus

Procyonidae (raccoon family)

Bassaricyon gabbi

Mustelidae (weasels)

Mustela africana

Felidae (cats)

Herpailurus yaguarondi

Cervidae (deer)

Mazama gouazoubira

Dinomyidae (pacarana)

Dinomys branickii

[1] For cats, we follow the taxonomy in Wilson and Reeder (1992)

Mammals of the Explorer's Inn Reserve

Louise H. Emmons, Linda J. Barkley and Mónica Romo

The following list includes mammals identified in the Explorer's Inn Reserve. The list is compiled chiefly from the following collections: (1) by Barkley in 1979 (largely bats); (2) by Emmons in Dec 1979-Jan 1980 (largely non-flying mammals); and (3) by Romo in 1985 (only bats). Bats, rodents and small marsupials are all vouchered by specimens in the Museo de Historia Natural, Lima; Louisiana State University Museum of Natural Science, Baton Rouge; or the United States National Museum of Natural History, Washington. Large mammals and some arboreal species are sight records. Some reliable records by other observers are included, but we omit unverified, less likely, anecdotal reports.

Didelphidae (opossums)

Caluromys lanatus

Marmosops noctivagus

Micoureus regina

Metachirus nudicaudatus

Philander opossum

Bradypodidae (three-toed sloths)

Bradypus variegatus

Megalonychidae (two-toed sloths)

Choloepus cf. *hoffmanni*

Dasypodidae (armadillos)

Dasypus kappleri

Dasypus novemcinctus

Priodontes maximus

Myrmecophagidae (anteaters)

Cyclopes didactylus

Myrmecophaga tridactyla

Tamandua tetradactyla

Emballonuridae (sheath-tailed bats)

Rhynchonycteris naso

Saccopteryx bilineata

Noctilionidae (fishing bats)

Noctilio albiventris

Noctilio leporinus

Phyllostomidae (leaf-nosed bats)

Artibeus anderseni

Artibeus jamaicencis

Artibeus lituratus

Artibeus obscurus

Artibeus sp.

Carollia brevicauda

Carollia castanea

Carollia perspicillata

Chiroderma salvini

Chiroderma villosum

Chrotopterus auritus

Desmodus rotundus

Glossophaga soricina

Lonchophylla thomasi

Mesophylla macconnelli

Micronycteris minuta

Phyllostomus elongatus

Phyllostomus hastatus

Platyrrhinus helleri

Rhinophylla pumilio

Sturnira lilium

Sturnira tildae

Tonatia bidens

Tonatia silvicola

Trachops cirrhosus

Uroderma bilobatum

Uroderma magnirostrum

Vampyrodes caraccioloi

Thyropteridae (sucker-footed bats)

Thyroptera tricolor

Vespertillionidae (vespertilionid bats)

Lasiurus ega

Myotis albescens

Myotis nigricans

Myotis riparius

Molossidae (free-tailed bats)

Tadarida brasiliensis

Callithricidae (tamarins)

Saguinus fuscicollis

Cebidae (monkeys)

Alouatta seniculus

Aotus trivirgatus

Callicebus moloch

Cebus apella

Cebus albifrons ?

Saimiri sciureus

Canidae (dogs)

Atelocynus microtis

Speothos venaticus

Procyonidae (raccoon family)

Bassaricyon alleni

Nasua nasua

Potos flavus

Procyon cancrivorus

Mustelidae (weasel family)

Eira barbara

Galictis vittata

Pteronura brasiliensis

Felidae (cats)

Herpailurus yaguarondi

Leopardus pardalis

Panthera onca

Puma concolor

Tapiridae (tapirs)

Tapirus terrestris

Tayassuidae (peccaries)

Tayassu pecari

Tayassu tajacu

Cervidae (deer)

Mazama americana

Mazama gouazoubira

Sciuridae (squirrels)

Sciurus ignitus

Sciurus spadiceus

Muridae (mice)

Nectomys squamipes

Oecomys bicolor

Oligoryzomys microtis

Oryzomys capito

Erethizontidae (porcupines)

Coendou cf. *bicolor*

Hydrochaeridae (capybara)

Hydrochaeris hydrochaeris

Agoutidae (pacas)

Agouti paca

Dasyproctidae (agoutis)

Dasyprocta variegata

Myoprocta pratti

Echimyidae (spiny rats)

Dactylomys dactylinus boliviensis

Mesomys hispidus

Proechimys brevicauda

Proechimys simonsi

Leporidae (rabbits)

Sylvilagus brasiliensis

Mammals of the Río Heath and Peruvian Pampas

Louise H. Emmons, César Ascorra and Mónica Romo

A. Species identified by Ascorra and first expedition (1); Emmons and Romo and second expedition (2); specimen in Louisiana State University Museum of Natural Science (3; L. Barkley and T. Parker, pers. comm.); or cited in Hofmann et al. 1976 (4). Species seen along the river are assigned to nearest camp locality. Localities are numbered as in gazetteer, except LSUMNS specimens from general area of Refugio Juliaca.

Locality:	6	7	8	9	10	In Pampa
Didelphidae (opossums)						
Didelphis marsupialis	1			1	2	
Caluromys lanatus	1,2					
Marmosops noctivagus					1*	
Micoureus regina			1*			
Philander opossum	1					
Myrmecophagidae (anteaters)						
Myrmecophaga tridactyla				2		4
Dasypodidae (armadillos)						
Dasypus sp.			1			
Priodontes maximus			1	1		
Emballonuridae (sheath-tailed bats)						
Rhynchonycteris naso				1*		
Noctilionidae (fishing bats)						
Noctilio albiventris	1		1	1		
Phyllostomidae (leaf-nosed bats)						
Artibeus gnomus					2*	1,2*
Artibeus jamaicencis	1*		1*			
Artibeus lituratus	1				2*,3	
Artibeus obscurus	1*		1*		2*,3	
Carollia castanea	1*		1*		1,3	
Carollia brevicauda	1*		1	1	1,2*,3	
Carollia perspicillata	1*		1*	1	1,2*,3	
Chiroderma villosum				1*		
Chrotopterus auritus			1*			
Glossophaga soricina	1*				3	
Lonchophylla thomasi	1*				2	2*
Mesophylla macconnelli			1*			
Micronycteris megalotis			1*		1*	

Locality:	6	7	8	9	10	In Pampa
Phyllostomus elongatus	1*					
Phyllostomus hastatus	1*					
Rhinophylla pumilio	1*					
Sturnira lilium			1*		3	
Sturnira magna	1*					
Sturnira tildae	1*					
Uroderma bilobatum	1*		1*			1*,3
Uroderma magnirostrum					3	
Tonatia silvicola			1*			
Trachops cirrhosus					3	
Vampyrodes caraccioli					3	
Vampyressa bidens					3	
Vespertilionidae (vesperilionid bats)						
Eptesicus brasiliensis thomasi					3	
Myotis nigricans					1*,3	
Myotis riparius			1*			
Molossidae (free-tailed bats)						
Molossus molossus					3	
Callithrichidae (tamarins)						
Saguinus fuscicollis	1				2	
Cebidae (monkeys)						
Aotus trivirgatus	1		1	1	2	
Alouatta seniculus	1,2		1	1	3	
Ateles paniscus	2					
Callicebus moloch	1			1,2	2	
Cebus apella	1,2	2	1	1,2	2	
Saimiri sciureus	2#		1	1		
Canidae (dogs)						
Atelocynus microtis				2		
Chrysocyon brachyurus					4	
Procyonidae (raccoon family)						
Nasua nasua			1	1,2,3		
Potos flavus	1			1		
Mustelidae (weasel family)						
Eira barbara				2		

#	On the Bolivian side
*	Specimen collected during this expedition

Locality:	6	7	8	9	10	In Pampa
Pteronura brasiliensis				1		
Felidae (cats)						
Leopardus wiedii	1		1	1		
Panthera onca			1	2	2	
Tapiridae (tapirs)						
Tapirus terrestris	1		1	1,2	2	1,2
Tayassuidae (peccaries)						
Tayassu tajacu	1,2			1,2		1
Tayassu pecari				2		
Cervidae (deer)						
Blastocerus dichotomus						1,2,4
Mazama americana	1		1	1,2		
Mazama gouazoubira		2?	1			
Sciuridae (squirrels)						
Sciurus ignitus	2					
Sciurus spadiceus	1,2			1,3		
Sciurus sanborni	1					
Muridae (rats)						
Bolomys lasiurus						2*
Oecomys bicolor					2*	
Oryzomys buccinatus						2*
Oryzomys capito			1			
Cavidae (guinea pigs)						
Cavia aperea						2*,3
Hydrochaeridae (capybara)						
Hydrochaeris hydrochaeris	1	1	1,2	1,2		
Dinomyidae (pacarana)						
Dinomys branickii	1					
Agoutidae (pacas)						
Agouti paca	1		1	1,2		
Dasyproctidae (agoutis)						
Dasyprocta variegata			1	1,2		
Echimyidae (spiny rats)						
Mesomys hispidus				1,2*		
Proechimys simonsi					2*	

CONSERVATION INTERNATIONAL

Rapid Assessment Program

B. Additional mammals reported to occur along the Río Heath by Darío Cruz Vani, a long-time resident, former professional hunter, and currently a park guard. These records require verification.

Myrmecophagidae

Tamandua tetradactyla

Bradypodidae

Bradypus variegatus

Megalonychidae

Choloepus sp.

Dasypodidae

Cabassous cf. *unicinctus*

Dasypus kappleri

Dasypus cf. *septemcinctus* (in pampa only)

Cebidae

Cebus albifrons

Canidae

Speothos venaticus

Procyonidae

Bassaricyon alleni

Mustelidae

Galictis vitttata

Lutra longicaudis

Felidae

Puma concolor

Herpailurus yaguarondi

Erethizontidae

Coendou cf. *bicolor*

Dasyproctidae

Dasyprocta - two spp. of different colors, yellow and black

Myoprocta pratti

Echimyidae

Dactylomys dactylinus

Leporidae

Sylvilagus brasiliensis

Amphibians and Reptiles in the Tambopata-Candamo Reserved Zone

Lily Rodríguez and Louise H. Emmons

	Ccolpa de Guacamayos (190m)	Cerros del Távara (250-900m)	Evidence
SALAMANDERS			
Plethodontidae			
Bolitoglossa sp.		1	-
FROGS			
Bufonidae			
Bufo marinus	T,R	T,R	x
Bufo poeppigi	T,R	T,R	x
Bufo sp. A,B,C (*typhonius* group)	T,F	T,R,S	x
Centrolenidae			
Cochranella sp.		1,S	x
Dendrobatidae			
Colostethus marchesianus		T,S	x
Dendrobates biolat	T,F		x
Epipedobates femoralis	T,F		x
Epipedobates pictus	T,F		x
Epipedobates trivittatus	T,F		x
Epipedobates sp. nov.		T,S	x
Hylidae			
Hemiphractus johnsoni		1,S	x
Hyla boans	A,R		x
Hyla calcarata	A,F		x
Hyla fasciata	1,F,p		x
Hyla geographica	1,R		x
Hyla lanciformis	1,R		x
Hyla leucophyllata	1,p		x
Hyla parviceps	1,F		x
Hyla sarayacuensis	1,F		x
Hyla callipleura		1,S,st	x
Scinax icterica	1,F		x
Scinax garbei	1,F		x
Scinax pedromedinae	1,F		x

	Ccolpa de Guacamayos (190m)	Cerros del Távara (250-900m)	Evidence
Osteocephalus leprieuri	A,F	A,S	x
Osteocephalus taurinus	A,F		x
Osteocephalus cf. *pearsoni*	A,F	A,S	x
Phyllomedusa bicolor	A,F,p	A	x
Phyllomedusa vaillanti	A,F	A	x
Phyllomedusa sp. nov.	A,F		x
Leptodactylidae			
Adenomera andreae	T,F		x
Adenomera hylaedactyla	T,O		c
Eleutherodactylus altamazonicus	l,F		x
Eleutherodactylus cf. *croceoinguinus*		l,F	x
Eleutherodactylus fenestratus	l,O	l,O	x
Eleutherodactylus imitatrix	l,F		x
Eleutherodactylus mendax	l,F		x
Eleutherodactylus ockendeni	l,F	l,S	x
Eleutherodactylus peruvianus	l,F	l,S	x
Eleutherodactylus toftae	l,F	l,S	x
Eleutherodactylus ventrimarmoratus	l,F		x
Eleutherodactylus sp. nov. a (78)	l,F	l,S	x
Eleutherodactylus sp. nov. b (92)		l,S	c
Eleutherodactylus sp. - yellow groin		l,S,st	x
Eleutherodactylus cf. *cruralis*	l,F,O		x
Ischnocnema quixensis		T,S	x
Leptodactylus knudseni	T,F		x
Leptodactylus leptodactyloides	T,F,O	T,R	x
Leptodactylus petersii	T,F		x
Leptodactylus pentadactylus	T,F		x
Leptodactylus rhodonotus *		T,S,st	x
Lithodytes lineatus	T,F		-
Physalaemus petersi	T,F		x
Phyllonastes myrmecoides *		T,R	x
Microhylidae			
Chiasmocleis bassleri	T,F		x
Hamptophryne boliviana	T,F,st		x

Habitat

T	Terrestrial
A	Arboreal
l	Low vegetation
O	Open area
p	Pond
R	River bank
st	Stream

Evidence

x	Specimen
f	Photograph
c	Call recorded
-	Sight
S	Species found in the ridges above the Ríos Guacamayo-Candamo junction
*****	Species collected by latter expedition

	Ccolpa de Guacamayos (190m)	Cerros del Távara (250-900m)	Evidence
LIZARDS			
Gekkonidae			
Gonatodes humeralis	I,F		-
Thecadactylus rapicauda	A,F		x
Hoplocercidae			
Enyaloides palpebralis	A,F		x
Polychridae			
Anolis fuscoauratus	I,F		x
Anolis punctatus	A,O		f
Scincidae			
Mabuya bistriata	I,O		-
Teiidae			
Ameiva ameiva	T,O		-
Cercosaura ocellata	T,F		x
Kentropyx altamazonica	T,F,O		-
Kentropyx pelviceps	I,F		-
Prionodactylus argulus	I,F		-
Tropiduridae			
Tropidurus umbra	A,F		x
SNAKES			
Boidae			
Boa constrictor	A		f
Corallus caninus	A		f
Corallus enydris	A		x
Epicrates cenchria	A	S,A	f
Colubridae			
Clelia clelia	T	S,st	x
Drymoluber dichrous	T		x
Imantodes cenchoa	A		x
Imantodes lentiferus	A		x
Leptodeira annulata	A		x
Tantilla melanocephala	T		x
Dendrophidion sp. nov.		S,I	x

	Ccolpa de Guacamayos (190m)	Cerros del Távara (250-900m)	Evidence
TURTLES			
Pelomedusidae			
Podocnemis unifilis	R		f
Testudinidae			
Geochelone denticulata	F		-
CAIMANS			
Alligatoridae			
Caiman crocodylus	R		-
Paleosuchus palpebrosus	st		f

Habitat

T	Terrestrial
A	Arboreal
l	Low vegetation
O	Open area
p	Pond
R	River bank
st	Stream

Evidence

x	Specimen
f	Photograph
c	Call recorded
-	Sight
S	Species found in the ridges above the Ríos Guacamayo-Candamo junction
*	Species collected by latter expedition

An asterisk (*) indicates species collected by J. Icochea on a latter expedition in September 1992 supported by TReeS and CONCYTEC (TReeS 1993; J. Icochea, pers. comm.).

Amphibians and Reptiles of the Pampas del Heath Region

Javier Icochea Monteza

	SAN	PIC	JUL	PAM	PPA	MIR
ANURA						
Bufonidae						
Bufo guttatus *		X				
Bufo marinus *	X	X	X			X
Bufo typhonius *			X			
Dendrobatidae						
Colostethus marchesianus *		X	X			
Hylidae						
Hyla calcarata *			X			
Hyla fasciata *	X	X				X
Hyla leucophyllata *		X				
Hyla sp. (albopunctata group) *				X		
Ololygon pedromedinai *	X					
Ololygon rubra *		X	X			X
Ololygon sp. *	X					
Osteocephalus taurinus *		X	X			
Osteocephalus cf. pearsoni *	X	X	X			
Phrynohyas venulosa *			X			
Phyllomedusa palliata *	X					
Leptodactylidae						
Adenomera andreae *		X				
Adenomera sp. *				X		
Eleutherodactylus fenestratus *	X	X	X		X	
Eleutherodactylus peruvianus *						X
Eleutherodactylus toftae *	X					
Leptodactylus mystaceus *			X			
Leptodactylus wagneri *	X	X	X			
Leptodactylus sp. *				X		
Physalaemus petersi *		X				
Microhylidae						
Hamptophryne boliviana *	X					
Pipidae						
Pipa pipa *			X			

CONSERVATION INTERNATIONAL

Rapid Assessment Program

	SAN	PIC	JUL	PAM	PPA	MIR
TESTUDINES						
Chelidae						
Phrynops gibbus *	X					
PELOMEDUSIDAE						
Podocnemis unifilis +	all along Río Heath					
SAURIA						
Gekkonidae						
Gonatodes humeralis *	X	X				
Polychridae						
Anolis fuscoauratus *			X			
Anolis punctatus *			X			
Scincidae						
Mabuya bistriata *			X			
Teiidae						
Ameiva ameiva +				X		
Kentropyx pelviceps *		X				
SERPENTES						
Boidae						
Corallus enydris *			X			
Eunectes murinus +				X		
Colubridae						
Atractus elaps *	X					
Helicops pastazae *		X				
Leptodeira annulata *		X				
CROCODYLIA						
Alligatoridae						
Caiman crocodilus +	all along Río Heath					
Paleosuchus palpebrosus *				X		

Localities

SAN	Puesto San Antonio
JUL	Refugio y Quebrada Juliaca
PIC	Refugio Picoplancha
PAM	Pampas del Heath (including headwaters of Quebrada Tapir)
PPA	Puerto Pardo
MIR	Fundo Miraflores

Evidence

+	Sight Record
*	Specimen

Fish Fauna of the Pampas del Heath National Sanctuary, Madre de Dios

Hernan Ortega

Number of fish species per family at six collecting localities.

Families	Collecting Stations 1	2	3	4	5	6
ENGRAULIDIDAE	0	0	1	2	0	0
CHARACIDAE	16	2	23	30	8	1
GASTEROPELECIDAE	2	0	0	2	3	0
ERYTHRINIDAE	1	0	1	2	0	1
LEBIASINIDAE	1	1	2	1	1	1
PARODONTIDAE	0	0	0	1	0	0
PROCHILODONTIDAE	0	0	0	1	0	0
CURIMATIDAE	0	0	3	4	0	0
ANOSTOMIDAE	0	0	2	3	0	0
GYMNOTIDAE	1	0	0	0	0	0
STERNOPYGIDAE	2	0	1	2	0	0
HYPOPOMIDAE	0	0	1	1	0	0
AUCHENIPTERIDAE	0	0	1	0	0	0
ASPREDINIDAE	1	0	1	0	0	0
PIMELODIDAE	5	0	2	5	0	0
TRICHOMYCTERIDAE	0	0	0	2	0	0
HELOGENIDAE	1	0	0	0	0	0
CALLICHTHYIDAE	2	1	0	0	1	0
LORICARIIDAE	1	0	0	5	0	0
BELONIDAE	0	0	1	0	0	0
RIVULIDAE	0	1	0	1	0	1
CICHLIDAE	1	0	2	0	1	0
SOLEIDAE	0	0	0	1	0	0
23 families	34	5	41	63	14	4

Collecting Stations

1 Quebrada San Antonio.—Mixture of white water and black water, mean width and depth 2.5 and 0.8 m respectively. Sandy-clay bottom, with leaf litter and detritus. Banks with shrubby and arborescent vegetation.

2 Seasonal ponds of San Antonio.—A series of three bodies of black water between 100 and 200 m² and 0.6 m deep, floating and rooted vegetation; clayey bottom with organic material of plant origin.

3 Cocha Picoplancha.—Lentic site 600 m long and 50 m wide; white water, but with a black water affluent. Bottom with decomposing leaf litter and sand.

4 Río Heath.—White water, lotic habitat, with a mean width of 50 m, depth of 1 to 4 m. Clayey-sand bottom and in parts with a hard bottom of alluvial origin, corresponding to the recent Holocene and which constitutes the southern extension of the Iñapari formation (Campbell et al. 1985).

5 Quebrada Juliaca.—White-black water creek, with a mean width of 8 m, 2.5 m depth and a clayey-sand bottom with organic material.

6 Quebrada Shuyo.—Clearwater creek, draining from the farthest end of the Pampa into the gallery forest. Mean width is 2 m and depth is 0.8 m. Clayey bottom with leaf litter and deposits of detritus in the pools.

Systematic List of the Fish Fauna of the Pampas del Heath National Sanctuary, Madre de Dios

Hernan Ortega

	Collecting Stations					
	1	2	3	4	5	6
ENGRAULIDIDAE						
Anchoviella sp.			X	X		
Lycengraulis sp.				X		
CHARACIDAE						
Aphiocharax sp. 1			X	X		
Aphiocharax sp. 2				X		
Astyanax bimaculatus	X		X	X	X	
Astyanax fasciatus			X	X		
Bario steindachneri	X					
Bryconella sp.				X		
Bryconops melanurus	X					
Bryconamericus sp.		X				
Characidium fasciatum	X		X			
Characidium sp.	X					
Charax pauciradiatus			X	X		
Cheirodon sp. 1	X			X		
Cheirodon sp. 2			X			
Creagrutus sp.				X		
Ctenobrycon hauxwellianus			X	X		
Cynopotamus amazonus				X		
Galeocharax gulo	X			X		
Gephyrocharax sp.	X			X		
Gymnocorymbus sp.			X	X		
Hemigrammus unilineatus		X		X	X	
Hyphessobrycon agulha					X	X
Hyphessobrycon copelandi			X	X		
Hyphessobrycon serpae			X	X		
Iguanodectes sp.				X		
Moenkhausia colleti	X		X	X	X	
Moenkhausia comma	X		X		X	
Moenkhausia dichroura			X	X		

	Collecting Stations					
	1	2	3	4	5	6
Moenkhausia doceana	X					
Moenkhausia oligolepis	X		X	X	X	
Moenkhausia sp.	X					
Moralesia sp.	X					
Phenacogaster sp.	X		X	X		
Piabina sp.	X		X	X	X	
Poptella sp.			X	X	X	
Prionobrama filigera			X	X		
Roeboides bicornis			X	X		
Roeboides sp.			X	X		
Salminus affinis				X		
Serrasalmus nattereri			X			
Serrasalmus rhombeus			X	X		
Tetragonopterus argenteus			X	X		
Triportheus sp.				X		
GASTEROPELECIDAE						
Carnegiella myersii	X				X	
Carnegiella strigata					X	
Gasteropelecus sp.	X			X	X	
Toracocharax stellatus				X		
ERYTHRINIDAE						
Hoplerythrinus unitaeniatus				X		X
Hoplias malabaricus	X		X	X		
LEBIASINIDAE						
Nannostomus trifasciatus			X			
Pyrrhulina brevis	X	X	X	X		
Pyrrhulina sp.					X	X
PARODONTIDAE						
Parodon sp.				X		
CURIMATIDAE						
Curimata sp.			X	X		
Potamorhina altamazonica						
Steindachnerina guentheri			X	X		
Steindachnerina hypostoma			X	X		

Collecting Stations	
1	Quebrada San Antonio
2	Seasonal ponds of San Antonio
3	Cocha Picoplancha
4	Río Heath
5	Quebrada Juliaca
6	Quebrada Shuyo

Collecting Stations

	1	2	3	4	5	6
Steindachnerina sp.				X		
PROCHILODONTIDAE						
Prochilodus nigricans				X		
ANOSTOMIDAE						
Leporinus arcus				X		
Leporinus friderici			X	X		
Leporinus striatus			X	X		
GYMNOTIDAE						
Gymnotus carapo	X					
HYPOPOMIDAE						
Hypopomus sp.			X	X		
STERNOPYGIDAE						
Eigenmannia humboldti	X					
Eigenmannia virescens	X		X	X		
Sternopygus sp.				X		
AUCHENIPTERIDAE						
Auchenipterus sp.			X			
ASPREDINIDAE						
Bunocephalus sp.	X		X			
PIMELODIDAE						
Duopalatinus goeldi				X		
Heptapterus sp.	X					
Nannorhamdia sp.	X					
Pimelodella sp. 1	X		X	X		
Pimelodella sp. 2	X		X	X		
Pimelodus maculatus				X		
Rhamdia sp.	X					
Sorubim lima				X		
TRICHOMYCTERIDAE						
Henonemus sp.				X		
Vandellia plazaii				X		
HELOGENIDAE						
Helogenes marmoratus	X					

Collecting Stations

	1	2	3	4	5	6
CALLICHTHYIDAE						
Callichthys sp.					X	
Corydoras stenocephalus	X					
Hoplosternum sp.	X	X				
LORICARIIDAE						
Hemiodontichthys acipenserinus	X			X		
Hypostomus sp.				X		
Planiloricaria sp.				X		
Pterigoplichthys multiradiatus				X		
Sturisoma sp.				X		
BELONIDAE						
Potamorhaphis sp.			X			
RIVULIDAE						
Pterolebias sp.		X				
Rivulus sp.				X		X
CICHLIDAE						
Apistogramma sp.			X			
Bujurquina sp.	X		X		X	
SOLEIDAE						
Aphionichthys sp.				X		
23 families	**34**	**5**	**41**	**63**	**14**	**4**

Collecting Stations

1	Quebrada San Antonio
2	Seasonal ponds of San Antonio
3	Cocha Picoplancha
4	Río Heath
5	Quebrada Juliaca
6	Quebrada Shuyo

List of Butterflies from Tambopata (Explorer's Inn Reserve)
Gerardo Lamas

NYMPHALIDAE
Heliconiinae
Acraeini

1. *Actinote pellenea hyalina* Jordan, 1913

Heliconiini

2. *Philaethria dido* (Linnaeus, 1763)
3. *Agraulis vanillae lucina* C & R Felder, 1862
4. *Dione juno juno* (Cramer, 1779)
5. *Dryadula phaetusa* (Linnaeus, 1758)
6. *Dryas iulia alcionea* (Cramer, 1779)
7. *Eueides aliphera aliphera* (Godart, 1819)
8. *E. heliconioides heliconioides* C & R Felder, 1861
9. *E. isabella hippolinus* Butler, 1873
10. *E. lampeto concisa* Lamas, 1985
11. *E. tales tabernula* Lamas, 1985
12. *E. vibilia unifasciata* Butler, 1873
13. *Laparus doris doris* (Linnaeus, 1771)
14. *Neruda aoede manu* (Lamas, 1976)
15. *Heliconius burneyi ada* Neustetter, 1925
16. *H. demeter tambopata* Lamas, 1985
17. *H. elevatus lapis* Lamas, 1976
18. *H. erato luscombei* Lamas, 1976
19. *H. hecale sisyphus* Salvin, 1871
20. *H. leucadia* Bates, 1862
21. *H. melpomene schunkei* Lamas, 1976
22. *H. numata lyrcaeus* Weymer, 1891
23. *H. pardalinus maeon* Weymer, 1891
24. *H. sara thamar* (Hübner, 1806)
25. *H. wallacei flavescens* Weymer, 1891

Nymphalinae
Nymphalini

26. *Hypanartia lethe* (Fabricius, 1793)

Kallimini

27. *Anartia amathea sticheli* Fruhstorfer, 1907
28. *A. jatrophae jatrophae* (Linnaeus, 1763)
29. *Metamorpha elissa elissa* Hübner, 1819
30. *Napeocles jucunda* (Hübner, 1808)
31. *Siproeta epaphus epaphus* (Latreille, 1813)
32. *S. stelenes meridionalis* (Fruhstorfer, 1909)

33. *Junonia genoveva occidentalis* C. & R. Felder, 1862

Melitaeini

34. *Anthanassa drusilla verena* (Hewitson, 1864)
35. *Castilia angusta* (Hewitson, 1868)
36. *C. perilla acraeina* (Hewitson, 1864)
37. *Mazia amazonica* ssp. n.
38. *Eresia clara clara* Bates, 1864
39. *E. eunice olivencia* Bates, 1864
40. *E. nauplius plagiata* (Röber, 1914)
41. *Ortilia gentina* Higgins, 1981
42. *Tegosa claudina* (Eschscholtz, 1821)
43. *T. serpia* Higgins, 1981
44. *Telenassa burchelli* (Moulton, 1909)

Limenitidinae
Coeini

45. *Historis acheronta acheronta* (Fabricius, 1775)
46. *H. odius* ssp. n.
47. *Baeotus amazonicus* (Riley, 1919)
48. *B. beotus* (Doubleday, 1849)
49. *B. deucalion* (C & R Felder, 1860)
50. *B. japetus* (Staudinger, 1885)
51. *Smyrna blomfildia blomfildia* (Fabricius, 1782)
52. *Colobura dirce dirce* (Linnaeus, 1758)
53. *Tigridia acesta tapajona* (Butler, 1873)

Biblidini

54. *Biblis hyperia laticlavia* (Thieme, 1904)
55. *Vila azeca azeca* (Doubleday, 1848)
56. *V. emilia caecilia* (C & R Felder, 1862)
57. *Myscelia capenas octomaculata* (Butler, 1873)
58. *Catonephele acontius acontius* (Linnaeus, 1771)
59. *C. antinoe* (Godart, 1824)
60. *C. numilia numilia* (Cramer, 1775)
61. *C. salacia* (Hewitson, 1852)
62. *Nessaea obrina lesoudieri* LeMoult, 1933
63. *Eunica alpais alpais* (Godart, 1824)
64. *E. amelia erroneata* Oberthür, 1916

65. *E. caelina alycia* Fruhstorfer, 1909

66. *E. clytia* (Hewitson, 1852)

67. *E. concordia* (Hewitson, 1852)

68. *E. eurota eurota* (Cramer, 1775)

69. *E. ingens* Seitz, 1916

70. *E. maja noerina* Hall, 1935

71. *E. malvina malvina* Bates, 1864

72. *E. marsolia fasula* Fruhstorfer, 1909

73. *E. mygdonia mygdonia* (Godart, 1824)

74. *E. norica occia* Fruhstorfer, 1909

75. *E. orphise* (Cramer, 1775)

76. *E. phasis* C & R Felder, 1862

77. *E. pusilla* Bates, 1864

78. *E. sophonisba agele* Seitz, 1916

79. *E. sydonia sydonia* (Godart, 1824)

80. *E. volumna celma* (Hewitson, 1852)

81. *Hamadryas amphinome amphinome* (Linnaeus, 1767)

82. *H. chloe chloe* (Stoll, 1787)

83. *H. feronia feronia* (Linnaeus, 1758)

84. *H. iphthime iphthime* (Bates, 1864)

85. *H. laodamia laodamia* (Cramer, 1777)

86. *Ectima iona* Doubleday, 1848

87. *E. thecla astricta* Fruhstorfer, 1908

88. *Panacea prola amazonica* Fruhstorfer, 1915

89. *P. regina* (Bates, 1864)

90. *Asterope markii hewitsoni* (Staudinger, 1886)

91. *Pyrrhogyra crameri hagnodorus* Fruhstorfer, 1908

92. *P. edocla cuparina* Bates, 1865

93. *P. neaerea amphiro* Bates, 1865

94. *P. otolais olivenca* Fruhstorfer, 1908

95. *Temenis laothoe laothoe* (Cramer, 1777)

96. *T. pulchra pallidior* (Oberthür, 1901)

97. *Nica flavilla sylvestris* Bates, 1864

98. *Peria lamis* (Cramer, 1779)

99. *Dynamine aerata aerata* (Butler, 1877)

100. *D. agacles agacles* (Dalman, 1823)

101. *D. artemisia glauce* (Bates, 1865)

102. *D. athemon barreiroi* Fernández, 1928

103. *D. coenus leucothea* (Bates, 1865)

104. *D. myrson* (Doubleday, 1849)

105. *D. postverta postverta* (Cramer, 1779)

106. *D. racidula racidula* (Hewitson, 1852)

107. *Haematera pyrame* ssp. n.

108. *Catacore kolyma pasithea* (Hewitson, 1864)

109. *Diaethria clymena peruviana* (Guenée, 1872)

110. *Paulogramma pyracmon peristera* (Hewitson, 1853)

111. *Callicore astarte stratiotes* (C & R Felder, 1861)

112. *C. cynosura cynosura* (Doubleday, 1847)

113. *C. eunomia carmen* (Oberthür, 1916)

114. *C. excelsior ockendeni* (Oberthür, 1916)

115. *C. hesperis* (Guérin, 1844)

116. *C. hystaspes zelphanta* (Hewitson, 1858)

117. *C. pygas cyllene* (Doubleday, 1847)

Limenitidini

118. *Adelpha aethalia davisii* (Butler, 1877)

119. *A. attica* (C & R Felder, 1867)

120. *A. cocala urraca* (C & R Felder, 1862)

121. *A. cytherea lanilla* Fruhstorfer, 1913

122. *A. epione agilla* Fruhstorfer, 1907

123. *A. iphiclus iphiclus* (Linnaeus, 1758)

124. *A. ixia pseudomessana* Fruhstorfer, 1913

125. *A. jordani* Fruhstorfer, 1913

126. *A. lerna lerna* (Hewitson, 1847)

127. *A. malea fugela* Fruhstorfer, 1916

128. *A. melona melona* (Hewitson, 1847)

129. *A. mesentina chancha* Staudinger, 1886

130. *A. naxia naxia* (C & R Felder, 1867)

131. *A. paraena* (Bates, 1865)

132. *A. phylaca juruana* (Butler, 1877)

133. *A. plesaure phliassa* (Godart, 1824)

134. *A. thesprotia delphicola* Fruhstorfer, 1909

135. *A. zunilaces* (?) ssp. n.

Cyrestidini

136. *Marpesia berania berania* (Hewitson, 1852)

137. *M. chiron marius* (Cramer, 1779)

138. *M. crethon* (Fabricius, 1776)

139. *M. egina* (Bates, 1865)

140. *M. furcula oechalia* (Westwood, 1850)

(SR) Sight record only

141. *M. petreus petreus* (Cramer, 1776)

142. *M. themistocles norica* (Hewitson, 1852)

Charaxinae

143. *Consul fabius divisus* (Butler, 1874)

144. *Hypna clytemnestra negra*
C. & R. Felder, 1862 (SR)

145. *Siderone galanthis thebais*
C. & R. Felder, 1862

146. *Zaretis itus itus* (Cramer, 1777)

147. *Fountainea chrysophana* (Bates, 1866)

148. *F. eurypyle eurypyle* (C & R Felder, 1862)

149. *Memphis basilia drucei* (Staudinger, 1887)

150. *M. cambyses* (Druce, 1877)

151. *M. glauce glauce* (C & R Felder, 1862)

152. *M. moruus memphis* (C & R Felder, 1867)

153. *M. phantes phantes* (Hopffer, 1874)

154. *M. phila morpheus* (Staudinger, 1886)

155. *M. philumena philumena* (Doubleday, 1849)

156. *M. polycarmes* (Fabricius, 1775)

157. *M. polyxo* (Druce, 1874)

158. *M. xenocles xenocles* (Hewitson, 1850)

159. *Archaeoprepona amphimachus symaithus*
Fruhstorfer, 1916

160. *A. demophon muson* (Fruhstorfer, 1905)

161. *A. demophoon andicola* (Fruhstorfer, 1904)

162. *A. licomedes* (Cramer, 1777)

163. *A. meander megabates* Fruhstorfer, 1916

164. *Prepona dexamenus dexamenus*
Hopffer, 1874

165. *P. laertes demodice* (Godart, 1824)

166. *P. pheridamas* (Cramer, 1777)

167. *P. pylene eugenes* Bates, 1865

168. *Agrias amydon aristoxenus* Niepelt, 1913

169. *A. claudina sardanapalus* Bates, 1860

Apaturinae

170. *Doxocopa agathina agathina* (Cramer, 1777)

171. *D. laure griseldis* (C & R Felder, 1862)

172. *D. lavinia* (Butler, 1866)

173. *D. linda linda* (C & R Felder, 1862)

174. *D. pavon pavon* (Latreille, 1809)

175. *D. zunilda floris* (Fruhstorfer, 1907)

Morphinae

176. *Antirrhea hela* C & R Felder, 1862

177. *A. philoctetes avernus* Hopffer, 1874

178. *A. taygetina taygetina* (Butler, 1868)

179. *Caerois chorinaeus protonoe*
Fruhstorfer, 1912

180. *Morpho absoloni* May, 1924

181. *M. achilles theodorus* Fruhstorfer, 1907

182. *M. deidamia grambergi* Weber, 1944

183. *M. hecuba cisseistricta*
LeMoult & Réal, 1962 (SR)

184. *M. menelaus alexandrovna* Druce, 1874

185. *M. telemachus iphiclus*
C & R Felder, 1862 (SR)

186. *M. zephyritis* Butler, 1873

Brassolinae

187. *Brassolis sophorae ardens* Stichel, 1903

188. *Narope cyllabarus* Westwood, 1851

189. *N. nesope* Hewitson, 1869

190. *N. panniculus* Stichel, 1904

191. *N. syllabus* Staudinger, 1887

192. *Dynastor darius* ssp. (SR)

193. *Opsiphanes cassiae crameri*
C & R Felder, 1862

194. *O. invirae amplificatus* Stichel, 1904

195. *O. quiteria quaestor* Stichel, 1902

196. *Opoptera aorsa hilara* Stichel, 1902

197. *Catoblepia berecynthia adjecta* Stichel, 1906

198. *C. soranus* Westwood, 1851

199. *C. xanthicles belisar* Stichel, 1904

200. *Selenophanes cassiope mapiriensis*
Bristow, 1982

201. *Eryphanis automedon tristis*
Staudinger, 1887

202. *Caligopsis seleucida seleucida*
(Hewitson, 1877)

203. *Caligo euphorbus euphorbus*
(C & R Felder, 1862)

204. *C. eurilochus livius* Staudinger, 1886

205. *C. idomeneus idomenides* Fruhstorfer, 1903

206. *C. illioneus praxsiodus* Fruhstorfer, 1912

207. *C. placidianus* Staudinger, 1887

208. *C. teucer semicaerulea* Joicey & Kaye, 1917

Satyrinae

209. *Cithaerias pireta* ssp. n.
210. *Haetera piera* ssp. n.
211. *Pierella astyoche boliviana* Brown, 1948
212. *P. hortona albofasciata* Rosenberg & Talbot, 1914
213. *P. hyalina extincta* Weymer, 1910
214. *P. lamia chalybaea* Godman, 1905
215. *P. lena brasiliensis* (C & R Felder, 1862)
216. *Bia actorion rebeli* Bryk, 1953
217. *Manataria hercyna hyrnethia* Fruhstorfer, 1912
218. *Harjesia blanda* (Möschler, 1877)
219. *H. obscura* (Butler, 1867)
220. *H. oreba* (Butler, 1870)
221. *Pseudodebis griseola* (Weymer, 1911)
222. *P. marpessa* (Hewitson, 1862)
223. *P. valentina* (Cramer, 1779)
224. *Taygetis angulosa* Weymer, 1907
225. *T. cleopatra* C & R Felder, 1862
226. *T. echo koepckei* Forster, 1964
227. *T. larua* C & R Felder, 1867
228. *T. mermeria mermeria* (Cramer, 1776)
229. *T. sosis* Hopffer, 1874
230. *T. sylvia* Bates, 1866
231. *T. thamyra* (Cramer, 1779)
232. *T. virgilia* (Cramer, 1776)
233. *Caeruleuptychia aegrota* (Butler, 1867)
234. *C. cyanites* (Butler, 1871)
235. *C. helios* (Weymer, 1911)
236. *C. lobelia* (Butler, 1870)
237. *C. scopulata* (Godman, 1905)
238. *C. ziza* (Butler, 1869)
239. *Cepheuptychia cephus cephus* (Fabricius, 1775)
240. *Cepheuptychia* sp. n.
241. *Chloreuptychia arnaca* (Fabricius, 1776)
242. *C. catharina* (Staudinger, 1886)
243. *C. herseis* (Godart, 1824)
244. *C. hewitsonii* (Butler, 1867)
245. *Chloreuptychia* sp. n. 1
246. *Chloreuptychia* sp. n. 2
247. *Cissia myncea* (Cramer, 1780)

248. *C. palladia* (Butler, 1867)
249. *C. penelope* (Fabricius, 1775)
250. *C. proba* (Weymer, 1911)
251. *C. terrestris* (Butler, 1867)
252. *Erichthodes erichtho* (Butler, 1867)
253. *Euptychia enyo* Butler, 1867
254. *Hermeuptychia hermes* (Fabricius, 1775)
255. *Magneuptychia aliciae* (Hayward, 1957)
256. *M. ayaya* (Butler, 1867)
257. *M. batesii* (Butler, 1867)
258. *M. gera nobilis* (Weymer, 1911)
259. *M. "helle"* (Cramer, 1779) - homonym
260. *M. lea philippa* (Butler, 1867)
261. *M. libye* (Linnaeus, 1767)
262. *M. moderata* (Weymer, 1911)
263. *M. modesta* (Butler, 1867)
264. *M. ocypete* (Fabricius, 1776)
265. *M. pallema* (Schaus, 1902)
266. *M. tricolor fulgora* (Butler, 1869)
267. *Magneuptychia* sp. n. 1
268. *Magneuptychia* sp. n. 2
269. *Magneuptychia* sp. n. 3
270. *Megeuptychia antonoe* (Cramer, 1775)
271. *Pareuptychia binocula binocula* (Butler, 1869)
272. *P. interjecta hesionides* Forster, 1964
273. *P. ocirrhoe* (Fabricius, 1776)
274. *P. summandosa* (Gosse, 1880)
275. *Paryphthimoides binalinea* (Butler, 1867)
276. *Posttaygetis penelea penelea* (Cramer, 1777)
277. *Rareuptychia clio* (Weymer, 1911)
278. *Splendeuptychia boliviensis* Forster, 1964
279. *S. furina* (Hewitson, 1862)
280. *S. itonis* (Hewitson, 1862)
281. *S. purusana* (Aurivillius, 1929)
282. *S. triangula* (Aurivillius, 1929)
283. *Splendeuptychia* sp. n. 1
284. *Splendeuptychia* sp. n. 2
285. *Splendeuptychia* sp. n. 3
286. *Splendeuptychia* sp. n. 4
287. *Yphthimoides mythra* (Weymer, 1911)
288. *Zischkaia amalda* (Weymer, 1911)

(SR) Sight record only

289. *"Euptychia" ordinata* (Weymer, 1911)
290. *Amphidecta calliomma* (C & R Felder, 1862)
291. *A. pignerator pignerator* Butler, 1867

Danainae

292. *Lycorea ilione phenarete* (Doubleday, 1847)
293. *L. halia pales* C & R Felder, 1862
294. *Danaus eresimus plexaure* (Godart, 1819)

Ithomiinae

295. *Athyrtis mechanitis salvini* Srnka, 1884
296. *Tithorea harmonia brunnea* Haensch, 1905
297. *Melinaea maelus lamasi* Brown, 1977
298. *M. marsaeus clara* Rosenberg & Talbot, 1914
299. *M. menophilus orestes* Salvin, 1871
300. *Paititia neglecta* Lamas, 1979
301. *Thyridia psidii ino* C & R Felder, 1862
302. *Sais rosalia badia* Haensch, 1905
303. *Forbestra olivencia aeneola* Fox, 1967
304. *Mechanitis lysimnia menecles* Hewitson, 1860
305. *M. mazaeus mazaeus* Hewitson, 1860
306. *M. polymnia angustifascia* Talbot, 1928
307. *Scada reckia labyrintha* Lamas, 1985
308. *Methona confusa psamathe* Godman & Salvin, 1898
309. *M. curvifascia* Weymer, 1883
310. *Rhodussa cantobrica pamina* (Haensch, 1905)
311. *Napeogenes aethra deucalion* Haensch, 1905
312. *N. inachia patientia* Lamas, 1985
313. *N. pharo pharo* (C & R Felder, 1862)
314. *Hypothyris euclea* ssp. n.
315. *Oleria didymaea didymaea* (Hewitson, 1876)
316. *O. onega lentita* Lamas, 1985
317. *O. ramona calatha* Lamas, 1985
318. *O. victorine victorine* (Guérin, 1844)
319. *Ithomia agnosia agnosia* Hewitson, 1855
320. *I. arduinna arduinna* d'Almeida, 1952
321. *I. lichyi neivai* d'Almeida, 1940
322. *I. salapia ardea* Hewitson, 1855
323. *Callithomia alexirrhoe thornax* Bates, 1862
324. *C. lenea zelie* (Guérin, 1844)

325. *Dircenna dero* ssp. n.
326. *D. loreta acreana* d'Almeida, 1950
327. *Ceratinia neso peruensis* (Haensch, 1905)
328. *C. tutia fuscens* (Haensch, 1905)
329. *Ceratiscada hymen hymen* (Haensch, 1905)
330. *Episcada polita angelita* Lamas, 1985
331. *Pteronymia antisao guntheri* Lamas, 1985
332. *P. forsteri* Baumann, 1985
333. *Hypoleria virginia vitiosa* Lamas, 1985
334. *"Hypoleria" orolina arzalia* (Hewitson, 1876)
335. *Pseudoscada timna* ssp. n.
336. *Heterosais nephele nephele* (Bates, 1862)

Libytheinae

337. *Libytheana carinenta carinenta* (Cramer, 1777)

RIODINIDAE

338. *Euselasia ignita* Stichel, 1924
339. *E. euboea euboea* (Hewitson, 1853)
340. *E. pellonia* Stichel, 1919
341. *E. eumedia eumedia* (Hewitson, 1853)
342. *E. mirania* (Bates, 1868)
343. *E. toppini* Sharpe, 1915
344. *Euselasia* aff. *phelina* (Druce, 1878)
345. *E. eusepus* (Hewitson, 1853)
346. *E. nannothis* Stichel, 1924
347. *E. euryone euryone* (Hewitson, 1856)
348. *E. violetta* (Bates, 1868)
349. *E. arbas* ssp.
350. *E. praecipua* Stichel, 1924
351. *E. euoras* (Hewitson, 1855)
352. *E. eutychus* (Hewitson, 1856)
353. *E. jugata* Stichel, 1919
354. *E. euodias euodias* (Hewitson, 1856)
355. *E. orba spectralis* Stichel, 1919
356. *E. issoria* (Hewitson, 1869)
357. *E. euriteus euriteus* (Cramer, 1777)
358. *E. eutaea eutaea* (Hewitson, 1853)
359. *E. zena* (Hewitson, 1860)
360. *E. opalina* (Westwood, 1851)
361. *E. praeclara* (Hewitson, 1869)
362. *E. melaphaea condensa* Stichel, 1927

363. *E. hygenius* group, sp. 1

364. *E. hygenius* group, sp. 2

365. *E. hygenius* group, sp. 3

366. *E. hygenius* group, sp. 4

367. *E. hygenius* group, sp. 5

368. *E. cafusa janigena* Stichel, 1919

369. *Euselasia* aff. *cafusa* (Bates, 1868)

370. *E. crinon crinon* Stichel, 1919

371. *Euselasia* sp. (*sergia rhodon* Seitz, 1917 ?)

372. *E. julia* (Druce, 1878) (?)

373. *E. labdacus reducta* Lathy, 1926

374. *E. gelanor erilis* Stichel, 1919

375. *E. murina* Stichel, 1925

376. *E. teleclus* (Stoll, 1787)

377. *Euselasia* sp., *midas* group

378. *E. pelusia* Stichel, 1919

379. *E. gordios* Stichel, 1919

380. *E. eugeon* (Hewitson, 1856)

381. *E. brevicauda* Lathy, 1926

382. *E. uria angustifascia* Lathy, 1926

383. *E. eubotes eubotes* (Hewitson, 1856)

384. *E. angulata* (Bates, 1868)

385. *E. utica euphaes* (Hewitson, 1855)

386. *Euselasia* aff. *eubotes* Hewitson, 1856

387. *Perophthalma tullius tullius* (Fabricius, 1787)

388. *Mesophthalma idotea* ssp. (n.?)

389. *Leucochimona hyphea pallida* (Lathy, 1932)

390. *L. mathata chionea*
    (Godman & Salvin, 1885)

391. *L. matisca* (Hewitson, 1860)

392. *Semomesia capanea sodalis* Stichel, 1919

393. *S. croesus siccata* Stichel, 1919

394. *S. macaris* (Hewitson, 1859)

395. *S. marisa marisa* (Hewitson, 1858)

396. *S. tenella tenella* Stichel, 1910

397. *Mesosemia* aff. *ephyne* (Cramer, 1776)

398. *Mesosemia* aff. *metura* Hewitson, 1873

399. *M. maeotis* Hewitson, 1859

400. *M. calypso calypso* Bates, 1868

401. *M. hesperina lycorias* Stichel, 1915

402. *M. cippus* Hewitson, 1859

403. *M. ibycus* Hewitson, 1859

404. *M. luperca* Stichel, 1910

405. *M. levis* Stichel, 1915

406. *M. olivencia olivencia* Bates, 1868

407. *M. philocles* ssp. n.

408. *M. machaera machaera* Hewitson, 1860

409. *M. thymetus umbrosa* Stichel, 1909

410. *M. subtilis* Stichel, 1909

411. *M. materna* Stichel, 1909

412. *Mesosemia* aff. *materna* Stichel, 1909

413. *M. naiadella naiadella* Stichel, 1909

414. *M. nerine* Stichel, 1909

415. *M. sirenia sirenia* Stichel, 1909

416. *M. judicialis* Butler, 1874

417. *Mesosemia* aff. *evias* Stichel, 1923

418. *M. menoetes paetula* Stichel, 1915

419. *M. ulrica ulrica* (Cramer, 1777)

420. *M. decolorata* Lathy, 1932

421. *Mesosemia* aff. *impedita* Stichel, 1909

422. *Eurybia nicaea* ssp.

423. *E. caerulescens caerulescens* Druce, 1904

424. *E. dardus franciscana* C & R Felder, 1862

425. *E. promota* Stichel, 1910

426. *E. halimede halimede* (Hübner, 1807)

427. *Alesa amesis* (Cramer, 1777)

428. *A. hemiurga* Bates, 1867

429. *Hyphilaria nicia* Hübner, 1819

430. *H. parthenis tigrinella* Stichel, 1909

431. *Napaea eucharila eucharila* (Bates, 1867)

432. *N. melampia* ssp.

433. *N. nepos nepos* (Fabricius, 1793)

434. *Cremna actoris meleagris* Hopffer, 1874

435. *C. thasus subrutila* Stichel, 1910

436. *Eunogyra satyrus* Westwood, 1851

437. *Lyropteryx apollonia apollonia*
    Westwood, 1851

438. *Cyrenia martia martia* Westwood, 1851

439. *Ancyluris meliboeus meliboeus*
    (Fabricius, 1776)

440. *A. etias melior* Stichel, 1910

441. *A. aulestes eryxo* (Saunders, 1859)

442. *Rhetus arcius huanus* (Saunders, 1859)

443. *R. periander laonome* (Morisse, 1838)

(SR) Sight record only

444. *Ithomeis lauronia* Schaus, 1902

445. *Isapis agyrtus sestus* (Stichel, 1909)

446. *Themone pais* ssp. n.

447. *Notheme erota diadema* Stichel, 1910

448. *Monethe albertus albertus*
    C & R Felder, 1862

449. *Metacharis lucius* (Fabricius, 1793)

450. *M. regalis regalis* Butler, 1867

451. *Cariomothis erythromelas fulvus* Lathy, 1932

452. *Pheles heliconides rufotincta* Bates, 1868

453. *Chamaelimnas tircis iaeris* Bates, 1868

454. *C. urbana* Stichel, 1916

455. *Parcella amarynthina* (C & R Felder, 1865)

456. *Charis anius* (Cramer, 1776)

457. *Charis* sp. n.

458. *C. gynaea zama* Bates, 1868

459. *Chalodeta theodora theodora*
    (C & R Felder, 1862)

460. *C. lypera* (Bates, 1868)

459. *C. chaonitis* (Hewitson, 1866)

462. *Caria mantinea amazonica* (Bates, 1868)

463. *C. trochilus arete* (C & R Felder, 1861).

464. *Comphotis drepana* (Bates, 1868)

465. *Crocozona coecias coecias* (Hewitson, 1866)

466. *Baeotis bacaenis bacaenita* Schaus, 1902

467. *Lasaia agesilas agesilas* (Latreille, 1809)

468. *L. arsis* Staudinger, 1887

469. *L. pseudomeris* Clench, 1972

470. *Amarynthis meneria* (Cramer, 1776)

471. *Exoplisia cadmeis* (Hewitson, 1866)

472. *Riodina lysippus lysias* Stichel, 1910

473. *Melanis xarifa quadripunctata* (Stichel, 1910)

474. *M. marathon stenotaenia* (Röber, 1904)

475. *Mesene leucophrys* Bates, 1868

476. *M. martha martha* Schaus, 1902

477. *M. epaphus sertata* Stichel, 1910

478. *M. nola eupteryx* Bates, 1868

479. *M. pyrrha* Bates, 1868

480. *Mesene* sp.

481. *M. monostigma* (Erichson, 1848) (?)

482. *Esthemopsis celina* Bates, 1868

483. *Symmachia* sp. 1
    (*?cleonyma* Hewitson, 1870)

484. *Symmachia* sp. 2 (*?probetor* Cramer, 1782)

485. *S. tricolor tricolor* Hewitson, 1867

486. *S. asclepia asclepia* Hewitson, 1870

487. *Phaenochitonia sophistes* (Bates, 1868)

488. *Sarota acantus* (Cramer, 1781)

489. *Sarota* sp. nr. *acantus* (Cramer, 1781)

490. *Sarota* aff. *myrtea* Godman & Salvin, 1886

491. *S. chrysus chrysus* (Cramer, 1781)

492. *Anteros formosus formosus* (Cramer, 1777)

493. *A. acheus troas* Stichel, 1909

494. *A. renaldus renaldus* (Stoll, 1791)

495. *Anteros* sp. n. 1 (*aerosus* Stichel, 1924?)

496. *Anteros* sp. n. 2

497. *Calydna punctata* C & R Felder, 1861

498. *C. catana* Hewitson, 1859

499. *C. maculosa* Bates, 1868

500. *C. euthria* Westwood, 1851

501. *Emesis castigata castigata* Stichel, 1910

502. *E. mandana mandana* (Cramer, 1780)

503. *E. fatimella fatimella* Westwood, 1851

504. *E. ocypore ocypore* (Geyer, 1837)

505. *E. cerea* (Linnaeus, 1767)

506. *E. temesa emesina* (Staudinger, 1887)

507. *Argyrogrammana occidentalis placibilis*
    (Stichel, 1910)

508. *A. physis phyton* (Stichel, 1911)

509. *A. praestigiosa* (Stichel, 1929) (?)

510. *A. trochilia rameli* (Stichel, 1930)

511. *Pachythone xanthe* Bates, 1868

512. *Callistium cleadas* (Hewitson, 1866)

513. *Uraneis hyalina* (Butler, 1867)

514. *Juditha azan* ssp. n.

515. *J. molpe molpe* (Hübner, 1808)

516. *Synargis orestessa* (Hübner, 1819)

517. *S. abaris* (Cramer, 1776)

518. *S. phylleus orontes* (Stichel, 1923)

519. *S. calyce calyce* (C & R Felder, 1862)

520. *S. gela gela* (Hewitson, 1853)

521. *S. agle* (Hewitson, 1853)

522. *S. ochra ochra* (Bates, 1868)

523. *S. regulus regulus* (Fabricius, 1793)

524. *Mycastor nealces* ssp. n.

525. *Parnes nycteis* Westwood, 1851

526. *P. philotes* Westwood, 1851

527. *Pandemos pasiphae* (Cramer, 1775)

528. *Calospila lucianus lucianus* (Fabricius, 1793)

529. *C. emylius emyliana* (Stichel, 1911)

530. *C. rhodope amphis* (Hewitson, 1870)

531. *C. parthaon* (Dalman, 1823)

532. *C. gyges* ssp.

533. *C. zeanger zeanger* (Stoll, 1790)

534. *C. rhesa* ssp.

535. *C. cerealis cerealis* (Hewitson, 1863)

536. *Adelotypa aristus* ssp.

537. *Adelotypa* aff. *aristus* (Stoll, 1790)

538. *A. huebneri pauxilla* (Stichel, 1911)

539. *A. mollis asemna* (Stichel, 1910)

540. *A. leucocyana* (Geyer, 1837)

541. *A. epixanthe* (Stichel, 1911)

542. *A. aminias aminias* (Hewitson, 1863)

543. *A. balista balista* (Hewitson, 1863)

544. *A. annulifera* (Godman, 1903)

545. *"Adelotypa" lampros* (Bates, 1868)

546. *Setabis epitus epiphanis* Stichel, 1910

547. *S. velutina* (Butler, 1867)

548. *S. lagus* group, sp. 1 (*pythioides*?)

549. *S. lagus* group, sp. 2

550. *S. lagus* group, sp. 3

551. *S. heliodora* (Staudinger, 1887)

552. *S. cruentata* (Butler, 1867)

553. *S. flammula* (Bates, 1868)

554. *Theope eudocia eudocia* Westwood, 1851

555. *Theope* sp. nr. *hypoleuca*
     Bates, 1868 (*foliorum*?)

556. *T. pedias pedias* Herrich-Schäffer, 1853

557. *T. sericea* Bates, 1868

558. *T. excelsa* Bates, 1868

559. *T. mundula* Stichel, 1926

560. *Theope* aff. *thootes* Hewitson, 1860

561. *Theope* aff. *thebais* ssp.

562. *Theope* aff. *thestias* Hewitson, 1860 (*discus*?)

563. *Nymphidium mantus* (Cramer, 1775)

564. *N. fulminans fulminans* Bates, 1868

565. *N. baeotia* Hewitson, 1853

566. *N. minuta* Druce, 1904

567. *N. azanoides amazonensis* Callaghan, 1986

568. *N. omois* Hewitson, 1865

569. *N. ascolia augea* Druce, 1904

570. *N. leucosia medusa* Druce, 1904

571. *N. acherois erymanthus* Ménétriès, 1855

572. *N. caricae parthenium* Stichel, 1924

573. *N. lisimon lisimon* (Stoll, 1790)

574. *N. hesperinum* Stichel, 1911

575. *Stalachtis calliope* ssp. n.

576. *S. phaedusa duvalii* (Perty, 1833)

LYCAENIDAE

577. *Hemiargus hanno* ssp.

578. *Mithras nautes* (Cramer, 1779)

579. *"Thecla"* nr. *orobia* (Hewitson, 1867)

580. *"Thecla" cosmophila* (Tessmann, 1928)

581. *"Thecla" maculata* (Lathy, 1936)

582. *Thestius meridionalis* (Draudt, 1921)

583. *"Thecla" ematheon* (Cramer, 1777)

584. *Evenus regalis* (Cramer, 1775)

585. *E. gabriela* (Cramer, 1775)

586. *E. batesii* (Hewitson, 1865)

587. *E. floralia* (Druce, 1907)

588. *E. satyroides* (Hewitson, 1865)

589. *"Thecla" falerina* (Hewitson, 1867)

590. *"Thecla" myrtea* (Hewitson, 1867)

591. *"Thecla" myrtusa* (Hewitson, 1867)

592. *Arcas actaeon* (Fabricius, 1775)

593. *A. tuneta* (Hewitson, 1865)

594. *Theritas mavors* Hübner, 1818

595. *Denivia acontius* (Goodson, 1945)

596. *D. phegeus* (Hewitson, 1865)

597. *D. viresco* (Druce, 1907)

598. *D. hemon* (Cramer, 1775)

599. *Denivia* nr. *lisus* (Stoll, 1790)

600. *Paiwarria telemus* (Cramer, 1775)

601. *"Thecla" ligurina* (Hewitson, 1874)

602. *"Thecla" ergina* (Hewitson, 1867)

603. *Thereus buris* (Druce, 1907)

604. *T. columbicola* (Strand, 1916)

605. *Arawacus separata* (Lathy, 1926)

606. *Rekoa palegon* (Cramer, 1780)

607. *Ocaria ocrisia* (Hewitson, 1868)

608. *Chlorostrymon telea* (Hewitson, 1868)

(SR) Sight record only

609. *Panthiades bitias* (Cramer, 1777)
610. *P. aeolus* (Fabricius, 1775)
611. *P. boreas* (C & R Felder, 1865)
612. *P. bathildis* (C & R Felder, 1865)
613. *P. phaleros* (Linnaeus, 1767)
614. *"Thecla" gemma* (Druce, 1907)
615. *"Thecla" minyia* (Hewitson, 1867)
616. *Parrhasius polibetes* (Cramer, 1781)
617. *P. orgia* (Hewitson, 1867)
618. *Michaelus ira* (Hewitson, 1867)
619. *M. vibidia* (Hewitson, 1869)
620. *M. thordesa* (Hewitson, 1867)
621. *"Thecla"* nr. *gadira* (Hewitson, 1867)
622. *"Thecla" norax* (Godman & Salvin, 1887)
623. *Olynthus* nr. *stigmatos* (Druce, 1890)
624. *O. essus* (Herrich-Schäffer, 1853)
625. *O. lorea* (Möschler, 1883)
626. *O. punctum* (Herrich-Schäffer, 1853)
627. *O. avoca* (Hewitson, 1867)
628. *Oenomaus atena* (Hewitson, 1867)
629. *Oenomaus* nr. *atena* (Hewitson, 1867) 1
630. *Oenomaus* nr. *atena* (Hewitson, 1867) 2
631. *Oenomaus* nr. *melleus* (Druce, 1907)
632. *Strymon daraba* (Hewitson, 1867)
633. *S. cestri* (Reakirt, 1867)
634. *S. ziba* (Hewitson, 1868)
635. *Lamprospilus orcidia* (Hewitson, 1874)
636. *L. aunus* (Cramer, 1775)
637. *L. genius* Geyer, 1832
638. *Lamprospilus* nr. *nisaee*
     (Godman & Salvin, 1887)
639. *L. lanckena* (Schaus, 1902)
640. *"Thecla" taminella* (Schaus, 1902)
641. *"Thecla"* nr. *galliena* (Hewitson, 1877)
642. *"Thecla" aruma* (Hewitson, 1877)
643. *"Thecla" syllis* (Godman & Salvin, 1887)
644. *"Thecla" hesperitis* (Butler & Druce, 1872)
645. *"Thecla" ceromia* (Hewitson, 1877)
646. *"Thecla"* nr. *ceromia* (Hewitson, 1877)
647. *"Thecla" vesper* (Druce, 1909)
648. *"Thecla" verbenaca* (Druce, 1907)
     (female only)

649. *Symbiopsis "peruviana"*
     (Lathy, 1936) - homonym
650. *Calycopis calus* (Godart, 1824)
651. *C. buphonia* (Hewitson, 1868)
652. *C. demonassa* (Hewitson, 1868)
653. *C. mimas* (Godman & Salvin, 1887)
654. *C. atnius* (Herrich-Schäffer, 1853)
655. *C. nicolayi* Field, 1967
656. *C. devia* (Möschler, 1883)
657. *C. vibulena* (Hewitson, 1877)
658. *C. vitruvia* (Hewitson, 1877)
659. *C. bellera* (Hewitson, 1877)
660. *C. anfracta* (Druce, 1907)
661. *C. anastasia* Field, 1967
662. *C. fractunda* Field, 1967
663. *C. centoripa* (Hewitson, 1868)
664. *C. caesaries* (Druce, 1907)
665. *C. cerata* (Hewitson, 1877)
666. *Calycopis* nr. *tifla* (Field, 1967)
667. *C. trebula* (Hewitson, 1868)
668. *Calycopis* nr. *vidulus* (Druce, 1907)
669. *C. orcillula* (Strand, 1916)
670. *C. atrox* (Butler, 1877)
671. *C. blora* (Field, 1967)
672. *Calycopis* nr. *naka* (Field, 1967)
673. *C. malta* (Schaus, 1902)
674. *C. anapa* (Field, 1967)
675. *Calycopis* nr. *barza* (Field, 1967)
676. *Tmolus echion* (Linnaeus, 1767)
677. *T. cydrara* (Hewitson, 1868)
678. *Tmolus* nr. *cydrara* (Hewitson, 1868)
679. *Tmolus* nr. *ufentina* (Hewitson, 1868)
680. *T. ufentina* (Hewitson, 1868)
681. *T. mutina* (Hewitson, 1867)
682. *"Thecla" gargophia* (Hewitson, 1877)
683. *"Thecla"* nr. *viceta* (Hewitson, 1868) 1
684. *"Thecla"* nr. *viceta* (Hewitson, 1868) 2
685. *"Thecla" emessa* (Hewitson, 1867)
686. *"Thecla"* nr. *tympania* (Hewitson, 1869) 1
687. *"Thecla"* nr. *tympania* (Hewitson, 1869) 2
688. *"Thecla"* nr. *tympania* (Hewitson, 1869) 3
689. *"Thecla" tympania* (Hewitson, 1869)
690. *"Thecla" tarena* (Hewitson, 1874)

691. *Siderus leucophaeus* (Hübner, 1813)

692. *S. parvinotus* Kaye, 1904

693. *Siderus* nr. *eliatha* (Hewitson, 1867)

694. *S. athymbra* (Hewitson, 1867)

695. *S. metanira* (Hewitson, 1867)

696. *S. philinna* (Hewitson, 1868)

697. *"Thecla" splendor* (Johnson, 1991)

698. *Theclopsis lydus* (Hübner, 1819)

699. *T. gargara* (Hewitson, 1868)

700. *"Thecla" tephraeus* (Geyer, 1837)

701. *"Thecla"* nr. *tephraeus* (Geyer, 1837)

702. *"Thecla" sphinx* (Fabricius, 1775)

703. *"Thecla" phoster* (Druce, 1907)

704. *"Thecla" pulchritudo* (Druce, 1907)

705. *"Thecla" strephon* (Fabricius, 1775)

706. *"Thecla"* nr. *strephon* (Fabricius, 1775)

707. *"Thecla" parvipuncta* (Lathy, 1926)

708. *"Thecla" agrippa* (Fabricius, 1793)

709. *"Thecla" carteia* (Hewitson, 1870)

710. *"Thecla"* nr. *carteia* (Hewitson, 1870)

711. *"Thecla" tyriam* (Druce, 1907)

712. *"Thecla"* nr. *tyriam* (Druce, 1907)

713. *"Thecla" syedra* (Hewitson, 1867)

714. *"Thecla"* nr. *syedra* (Hewitson, 1867)

715. *"Thecla" occidentalis* (Lathy, 1926)

716. *Ministrymon megacles* (Cramer, 1780)

717. *M. zilda* (Hewitson, 1873)

718. *M. cruenta* (Gosse, 1880)

719. *Ministrymon* nr. *cruenta* (Gosse, 1880)

720. *"Thecla" terentia* (Hewitson, 1868)

721. *Aubergina alda* (Hewitson, 1868)

722. *Janthecla rocena* (Hewitson, 1867)

723. *J. malvina* (Hewitson, 1867)

724. *J. leea* Venables & Robbins, 1991

725. *J. sista* (Hewitson, 1867)

726. *Hypostrymon asa* (Hewitson, 1873)

727. *Iaspis verania* (Hewitson, 1868)

728. *I. castitas* (Druce, 1907)

729. *Iaspis* nr. *castitas* (Druce, 1907)

730. *I. temesa* (Hewitson, 1868)

731. *"Thecla" picus* (Druce, 1907)

732. *Brangas getus* (Fabricius, 1787)

733. *"Thecla" cupentus* (Cramer, 1781)

734. *Trichonis immaculata* Lathy, 1930

735. *"Thecla" biston* (Möschler, 1877)

736. *"Thecla" bactriana* (Hewitson, 1868)

737. *Nesiostrymon celona* (Hewitson, 1874)

738. *Erora phrosine* (Druce, 1909)

739. *Chalybs hassan* (Stoll, 1790)

740. *"Thecla" tema* (Hewitson, 1867)

741. *Caerofethra carnica* (Hewitson, 1873)

742. *Celmia celmus* (Cramer, 1775)

743. *"Thecla" color* (Druce, 1907)

744. *"Thecla" mecrida* (Hewitson, 1867)

745. *"Thecla" conoveria* (Schaus, 1902)

746. *"Thecla" ceglusa* (Hewitson, 1868)

PIERIDAE

Dismorphiinae

747. *Pseudopieris nehemia melania* Lamas, 1985

748. *Dismorphia theucharila argochloe*
    (Bates, 1861)

749. *Enantia lina galanthis* (Bates, 1861)

750. *E. melite linealis* (Prüffer, 1922)

Coliadinae

751. *Anteqs menippe* (Hübner, 1818)

752. *Aphrissa fluminensis* (d'Almeida, 1921)

753. *A. statira statira* (Cramer, 1777)

754. *Phoebis argante larra* (Fabricius, 1798)

755. *P. philea philea* (Linnaeus, 1763)

756. *P. sennae marcellina* (Cramer, 1777) (SR)

757. *Rhabdodryas trite trite* (Linnaeus, 1758)

758. *Eurema agave agave* (Cramer, 1775)

759. *E. albula espinosae* (Fernández, 1928)

760. *E. elathea obsoleta* (Jörgensen, 1932)

761. *E. paulina* (Bates, 1861)

762. *Pyrisitia leuce flavilla* (Bates, 1861)

763. *P. nise* ssp. n.

Pierinae

764. *Hesperocharis emeris aida* Fruhstorfer, 1908

765. *Glutophrissa drusilla drusilla* (Cramer, 1777)

766. *Ascia monuste automate* (Burmeister, 1878)

767. *Ganyra phaloe sublineata* (Schaus, 1902)

768. *Itaballia demophile lucania*
    (Fruhstorfer, 1907)

(SR) Sight record only

769. *I. pandosia pisonis* (Hewitson, 1861)
770. *Pieriballia viardi rubecula* (Fruhstorfer, 1907)
771. *Melete lycimnia peruviana* (Lucas, 1852)
772. *Perrhybris pamela mazuka* Lamas, 1981

PAPILIONIDAE

Papilioninae

773. *Protographium agesilaus autosilaus* (Bates, 1861)
774. *Eurytides dolicaon deileon* (C & R Felder, 1865)
775. *Protesilaus glaucolaus leucas* (Rothschild & Jordan, 1906)
776. *P. molops hetaerius* (Rothschild & Jordan, 1906)
777. *P. telesilaus telesilaus* (C & R Felder, 1864)
778. *Mimoides ariarathes gayi* (Lucas, 1852)
779. *M. pausanias pausanias* (Hewitson, 1852)
780. *M. xynias xynias* (Hewitson, 1875)
781. *Battus belus belus* (Cramer, 1777)
782. *B. crassus* (Cramer, 1777)
783. *B. lycidas* (Cramer, 1777)
784. *B. polydamas polydamas* (Linnaeus, 1758)
785. *Parides aeneas lamasi* Racheli, 1988
786. *P. anchises drucei* (Butler, 1874)
787. *P. lysander brissonius* (Hübner, 1819)
788. *P. neophilus olivencius* (Bates, 1861)
789. *P. pizarro kuhlmanni* (May, 1925)
790. *P. sesostris sesostris* (Cramer, 1779)
791. *P. vertumnus astorius* (Zikán, 1940)
792. *Heraclides anchisiades anchisiades* (Esper, 1788)
793. *H. androgeus androgeus* (Cramer, 1775)
794. *H. chiansiades chiansiades* (Westwood, 1872)
795. *H. hyppason* (Cramer, 1775)
796. *H. thoas cinyras* (Ménétriès, 1857)
797. *H. torquatus torquatus* (Cramer, 1777)

HESPERIIDAE

Pyrrhopyginae

798. *Pyrrhopyge phidias bixae* (Linnaeus, 1758)
799. *P. aziza attis* Bell, 1931
800. *P. proculus draudti* Bell, 1931
801. *P. thericles rileyi* Bell, 1931
802. *P. amythaon perula* Evans, 1951
803. *P. sergius josephina* Draudt, 1921
804. *P. cometes* ssp. n.
805. *P. thelersa* Hewitson, 1866
806. *Elbella intersecta intersecta* (Herrich-Schäffer, 1869)
807. *E. merops* (Bell, 1934)
808. *E. patrobas* ssp. n.
809. *E. blanda* Evans, 1951
810. *E. azeta azeta* (Hewitson, 1866)
811. *Elbella* sp. n.
812. *Nosphistia zonara* (Hewitson, 1866)
813. *Jemadia hospita hospita* (Butler, 1877)
814. *J. hewitsonii hewitsonii* (Mabille, 1878)
815. *J. gnetus* (Fabricius, 1782)
816. *Mysoria sejanus* ssp. n.
817. *Myscelus nobilis* (Cramer, 1777)
818. *M. amystis mysus* Evans, 1951
819. *M. pardalina pardalina* (C. & R. Felder, 1867)
820. *Passova ganymedes gelina* Evans, 1951
821. *P. passova styx* (Möschler, 1879)
822. *Aspitha agenoria sanies* (Druce, 1908)
823. *Oxynetra confusa* Staudinger, 1888

Pyrginae

824. *Phocides distans distans* (Herrich-Schäffer, 1869)
825. *P. metrodorus metrodorus* Bell, 1932
826. *P. vulcanides* Röber, 1925
827. *P. novalis* Evans, 1952
828. *P. padrona* Evans, 1952
829. *P. pigmalion hewitsonius* (Mabille, 1883)
830. *Tarsoctenus papias* (Hewitson, 1859)
831. *T. corytus corba* Evans, 1952
832. *T. praecia plutia* (Hewitson, 1857)
833. *Phanus vitreus* (Cramer, 1781)

834. *P. ecitonorum* Austin, 1993

835. *P. marshalli* (Kirby, 1880)

836. *Udranomia kikkawai* (Weeks, 1906)

837. *Drephalys dumeril* (Latreille, 1824)

838. *D. oriander oriander* (Hewitson, 1867)

839. *D. olvina* Evans, 1952

840. *D. alcmon* (Cramer, 1779)

841. *D. hypargus* (Mabille, 1891)

842. *Augiades crinisus* (Cramer, 1780)

843. *Hyalothyrus leucomelas* (Geyer, 1832)

844. *Phareas coeleste* Westwood, 1852

845. *Entheus eumelus ninyas* Druce, 1912

846. *E. gentius* (Cramer, 1777)

847. *E. priassus telemus* Mabille, 1898

848. *Proteides mercurius mercurius*
    (Fabricius, 1787)

849. *Epargyreus socus sinus* Evans, 1952

850. *E. exadeus exadeus* (Cramer, 1779)

851. *E. clavicornis clavicornis*
    (Herrich-Schäffer, 1869)

852. *Polygonus manueli manueli*
    Bell & Comstock, 1948

853. *Chioides catillus catillus* (Cramer, 1779)

854. *Aguna* sp. n.

855. *A. aurunce* (Hewitson, 1867) (?)

856. *A. coelus* (Cramer, 1782) (?)

857. *Polythrix octomaculata octomaculata*
    (Sepp, 1844)

858. *P. minvanes* (Williams, 1926)

859. *P. auginus* (Hewitson, 1867) (?)

860. *P. metallescens* (Mabille, 1888)

861. *P. eudoxus* (Cramer, 1781)

862. *Heronia labriaris* (Butler, 1877)

863. *Chrysoplectrum pervivax* (Hübner, 1819)

864. *C. perniciosus perniciosus*
    (Herrich-Schäffer, 1869)

865. *Codatractus* sp. n.

866. *Urbanus pronta* Evans, 1952

867. *U. esmeraldus* (Butler, 1877)

868. *U. velinus* (Plötz, 1880)

869. *U. dorantes dorantes* (Stoll, 1790)

870. *U. teleus* (Hübner, 1821)

871. *U. simplicius* (Stoll, 1790)

872. *U. reductus* (Riley, 1919)

873. *U. doryssus doryssus* (Swainson, 1831)

874. *U. albimargo takuta* Evans, 1952

875. *U. virescens* (Mabille, 1877)

876. *U. chalco* (Hübner, 1823)

877. *Cephise cephise* (Herrich-Schäffer, 1869) (?)

878. *Astraptes talus* (Cramer, 1777)

879. *A. fulgerator fulgerator* (Walch, 1775)

880. *A. halesius* (Hewitson, 1877)

881. *A. apastus apastus* (Cramer, 1777)

882. *A. aulus* (Plötz, 1881)

883. *A. enotrus* (Cramer, 1782)

884. *A. janeira* (Schaus, 1902)

885. *A. alector hopfferi* (Plötz, 1882)

886. *A. cretatus cretatus* (Hayward, 1939)

887. *A. creteus creteus* (Cramer, 1780)

888. *A. chiriquensis* ssp. n.

889. *Narcosius narcosius narcosius* (Stoll, 1790)

890. *N. mura* (Williams, 1927)

891. *Calliades zeutus* (Möschler, 1879)

892. *Autochton neis* (Geyer, 1832)

893. *A. longipennis* (Plötz, 1882)

894. *A. zarex* (Hübner, 1818)

895. *Venada advena* (Mabille, 1889)

896. *Bungalotis astylos* (Cramer, 1780)

897. *B. borax lactos* Evans, 1952

898. *Salatis salatis* (Cramer, 1782)

899. *Sarmientoia eriopis* (Hewitson, 1867)

900. *Dyscophellus nicephorus* (Hewitson, 1876)

901. *D. euribates euribates* (Cramer, 1782)

902. *D. sebaldus* (Cramer, 1781)

903. *D. erythras* (Mabille, 1888)

904. *D. ramusis ramusis* (Cramer, 1781)

905. *Nascus phintias* Schaus, 1913

906. *N. solon solon* (Plötz, 1882)

907. *N. paulliniae* (Sepp, 1842)

908. *Porphyrogenes zohra stresa* Evans, 1952

909. *Oileides azines* (Hewitson, 1867)

910. *Celaenorrhinus shema shema*
    (Hewitson, 1877)

911. *C. disjunctus* Bell, 1940

912. *Celaenorrhinus* sp. (*similis* group)

913. *C. syllius* (C & R Felder, 1862)

(SR) Sight record only

914. *C. jao* (Mabille, 1889)

915. *Marela tamyris tamyris* Mabille, 1903

916. *Telemiades delalande* (Latreille, 1824)

917. *T. trenda* Evans, 1953

918. *T. epicalus epicalus* Hübner, 1819

919. *T. penidas* (Hewitson, 1867)

920. *T. avitus* (Cramer, 1781)

921. *T. antiope tosca* Evans, 1953

922. *T. amphion misitheus* Mabille, 1888

923. *Eracon paulinus* (Cramer, 1782)

924. *E. onoribo* (Möschler, 1883)

925. *Spioniades artemides* (Cramer, 1782)

926. *S. libethra* (Hewitson, 1868)

927. *Mictris crispus* (Herrich-Schäffer, 1870)

928. *Iliana heros heros* (Mabille & Boullet, 1917)

929. *Sophista aristoteles aristoteles*
(Westwood, 1852)

930. *Polyctor polyctor polyctor* (Prittwitz, 1868)

931. *Nisoniades bessus* (Möschler, 1877)

932. *N. mimas* (Cramer, 1775)

933. *N. evansi* Steinhauser, 1989

934. *N. brunneata* (Williams & Bell, 1939)

935. *N. macarius* Herrich-Schäffer, 1870

936. *Pachyneuria lineatopunctata lineatopunctata*
(Mabille & Boullet, 1917)

937. *P. duidae duidae* (Bell, 1932)

938. *P. herophile* (Hayward, 1940)

939. *Pellicia klugi* Williams & Bell, 1939

940. *P. santana* Williams & Bell, 1939

941. *P. costimacula costimacula*
Herrich-Schäffer, 1870

942. *P. trax* Evans, 1955

943. *Morvina fissimacula rema* Evans, 1953

944. *Cyclosemia earina* (Hewitson, 1878)

945. *C. herennius elelea* (Hewitson, 1878)

946. *Gorgopas trochilus* (Hopffer, 1874)

947. *Viola olla* Evans, 1953

948. *Bolla mancoi* (Lindsey, 1925)

949. *B. cupreiceps* (Mabille, 1891)

950. *B. morona morona* (Bell, 1940)

951. *Staphylus chlora* Evans, 1953

952. *S. putumayo* (Bell, 1937)

953. *S. lizeri lizeri* (Hayward, 1938)

954. *S. corumba* (Williams & Bell, 1940)

955. *S. oeta* (Plötz, 1884)

956. *S. astra* (Williams & Bell, 1940)

957. *Plumbago plumbago* (Plötz, 1884)

958. *Gorgythion begga pyralina* (Möschler, 1877)

959. *G. beggina escalophoides* Evans, 1953

960. *Ouleus juxta juxta* (Bell, 1934)

961. *O. fridericus fridericus* (Geyer, 1832)

962. *O. accedens noctis* (Lindsey, 1925)

963. *Quadrus cerialis* (Cramer, 1782)

964. *Q. contubernalis contubernalis*
(Mabille, 1883)

965. *Q. deyrollei porta* Evans, 1953

966. *Gindanes brebisson phagesia*
(Hewitson, 1868)

967. *Pythonides jovianus fabricii* Kirby, 1871

968. *P. lerina* (Hewitson, 1868)

969. *P. grandis assecla* Mabille, 1883

970. *P. herennius herennius* Geyer, 1838

971. *P. braga* Evans, 1953

972. *P. maraca* ssp. n.

973. *Sostrata festiva* (Erichson, 1848)

974. *S. pusilla pusilla* Godman & Salvin, 1895

975. *Paches trifasciatus* Lindsey, 1925

976. *Haemactis sanguinalis* (Westwood, 1852)

977. *Milanion hemes* ssp.

978. *M. hemestinus* Mabille & Boullet, 1917

979. *M. pilumnus pilumnus*
Mabille & Boullet, 1917

980. *Charidia lucaria lucaria* (Hewitson, 1868)

981. *Mylon ander ander* Evans, 1953

982. *M. menippus* (Fabricius, 1776)

983. *M. pelopidas* (Fabricius, 1793)

984. *M. jason* (Ehrmann, 1907)

985. *Carrhenes fuscescens conia* Evans, 1953

986. *C. canescens leada* (Butler, 1870)

987. *C. santes* Bell, 1940

988. *Clito clito* (Fabricius, 1787)

989. *Xenophanes tryxus* (Cramer, 1780)

990. *Antigonus nearchus* (Latreille, 1817)

991. *A. erosus* (Hübner, 1812)

992. *A. decens* Butler, 1874

993. *Anisochoria pedaliodina pedaliodina* (Butler, 1870)

994. *Aethilla echina echina* Hewitson, 1870

995. *Achlyodes busirus heros* Ehrmann, 1909

996. *A. mithridates thraso* (Hübner, 1807)

997. *Anastrus sempiternus simplicior* (Möschler, 1877)

998. *A. tolimus robigus* (Plötz, 1884)

999. *A. petius petius* (Möschler, 1877)

1000. *A. obliqua* (Plötz, 1884)

1001. *A. meliboea bactra* Evans, 1953

1002. *A. obscurus narva* Evans, 1953

1003. *Ebrietas infanda* (Butler, 1877)

1004. *E. anacreon anacreon* (Staudinger, 1876)

1005. *E. evanidus* Mabille, 1898

1006. *Cycloglypha thrasibulus thrasibulus* (Fabricius, 1793)

1007. *C. tisias* (Godman & Salvin, 1896)

1008. *Helias phalaenoides phalaenoides* (Hübner, 1812)

1009. *Camptopleura theramenes* Mabille, 1877

1010. *C. auxo* (Möschler, 1879)

1011. *C. impressus* (Mabille, 1889)

1012. *Chiomara mithrax* (Möschler, 1879)

1013. *Pyrgus oileus orcus* (Cramer, 1780)

1014. *Heliopetes alana* (Reakirt, 1868)

Hesperiinae

1015. *Synapte silius* (Latreille, 1824)

1016. *Lento* sp. n. 1

1017. *L. ferrago* (Plötz, 1884)

1018. *L. lucto* Evans, 1955

1019. *L. imerius* (Plötz, 1884)

1020. *L. lora* Evans, 1955

1021. *Lento* sp. n. 2

1022. *Anthoptus epictetus* (Fabricius, 1793)

1023. *A. insignis* (Plötz, 1882)

1024. *Corticea* sp. (n.?)

1025. *C. corticea* (Plötz, 1882)

1026. *Cantha zara* (Bell, 1941)

1027. *C. calva* Evans, 1955

1028. *Vinius sagitta* (Mabille, 1889)

1029. *V. tryhana tryhana* (Kaye, 1914)

1030. *Pheraeus maria* Steinhauser, 1991

1031. *Pheraeus* sp. n. 1

1032. *Pheraeus* sp. n. 2

1033. *Misius misius* (Mabille, 1891)

1034. *Molo mango* (Guenée, 1865)

1035. *M. calcarea* ssp. n.

1036. *Callimormus radiola radiola* (Mabille, 1878)

1037. *Radiatus bradus* Mielke, 1968

1038. *Eutocus matildae vinda* Evans, 1955

1039. *E. quichua* Lindsey, 1925

1040. *Virga virginius* (Möschler, 1883)

1041. *V. phola* Evans, 1955 (?)

1042. *Ludens silvaticus* (Hayward, 1940)

1043. *Methionopsis ina* (Plötz, 1882)

1044. *M. dolor* Evans, 1955

1045. *Sodalia sodalis* (Butler, 1877)

1046. *Artines aepitus* (Geyer, 1832)

1047. *A. focus* Evans, 1955

1048. *A. trogon* Evans, 1955

1049. *Artines* sp. n.

1050. *Flaccilla aecas* (Cramer, 1781)

1051. *Mnaseas bicolor inca* Bell, 1930

1052. *Gallio* sp. n.

1053. *Thargella caura caura* (Plötz, 1882)

1054. *Venas evans* (Butler, 1877)

1055. *V. caerulans* (Mabille, 1878)

1056. *Phanes aletes* (Geyer, 1832)

1057. *P. almoda* (Hewitson, 1866)

1058. *Vidius* sp. n.

1059. *Nastra guianae* (Lindsey, 1925)

1060. *Cymaenes tripunctus theogenis* (Capronnier, 1874)

1061. *C. hazarma* (Hewitson, 1877)

1062. *C. cavalla* Evans, 1955

1063. *C. uruba taberi* (Weeks, 1901)

1064. *Vehilius stictomenes stictomenes* (Butler, 1877)

1065. *V. inca* (Scudder, 1872)

1066. *V. almoneus* (Schaus, 1902)

1067. *V. vetulus* (Mabille, 1878)

1068. *V. putus* Bell, 1941

1069. *Vehilius madius* ssp. n.

1070. *Mnasilus allubita* (Butler, 1877)

1071. *Mnasitheus chrysophrys* (Mabille, 1891)

(SR) Sight record only

1072. *M. forma* Evans, 1955

1073. *M. nitra* Evans, 1955

1074. *Remella remus* (Fabricius, 1798)

1075. *Moeris submetallescens* ssp. n.

1076. *Parphorus storax storax* (Mabille, 1891)

1077. *P. decora* (Herrich-Schäffer, 1869)

1078. *Parphorus* sp. n. 1

1079. *Parphorus* sp. n. 2

1080. *Papias phainis* Godman, 1900

1081. *P. phaeomelas* (Hübner, 1831)

1082. *P. tristissimus* Schaus, 1902

1083. *Propapias proximus* (Bell, 1934)

1084. *Cobalopsis nero* (Herrich-Schäffer, 1869)

1085. *C. venias* (Bell, 1942)

1086. *C. dorpa* de Jong, 1983

1087. *Arita arita* (Schaus, 1902)

1088. *Morys compta compta* (Butler, 1877)

1089. *M. geisa geisa* (Möschler, 1879)

1090. *Morys* (?) sp.

1091. *Psoralis* sp. n.

1092. *Tigasis garima garima* (Schaus, 1902)

1093. *Tigasis* sp. n.

1094. *Vettius lafrenaye pica*
    (Herrich-Schäffer, 1869)

1095. *V. richardi* (Weeks, 1906)

1096. *V. triangularis* (Hübner, 1831)

1097. *V. monacha* (Plötz, 1882)

1098. *V. phyllus phyllus* (Cramer, 1777)

1099. *V. marcus marcus* (Fabricius, 1787)

1100. *V. artona* (Hewitson, 1868)

1101. *V. yalta* Evans, 1955

1102. *V. fuldai* (Bell, 1930)

1103. *Vettius* sp. n.

1104. *Paracarystus hypargyra*
    (Herrich-Schäffer, 1869)

1105. *P. menestries rona* (Hewitson, 1866)

1106. *Turesis complanula* (Herrich-Schäffer, 1869)

1107. *T. basta* Evans, 1955

1108. *Thoon canta* Evans, 1955

1109. *T. modius* (Mabille, 1889)

1110. *T. dubia* (Bell, 1932)

1111. *T. taxes* (Godman, 1900)

1112. *T. ponka* Evans, 1955

1113. *T. aethus* (Hayward, 1951)

1114. *T. ranka* Evans, 1955

1115. *Thoon* sp. n. (nr. *yesta* Evans, 1955)

1116. *Justinia phaetusa phaetusa* (Hewitson, 1866)

1117. *J. justinianus dappa* Evans, 1955

1118. *Justinia* sp. n.

1119. *Eutychide complana*
    (Herrich-Schäffer, 1869)

1120. *E. subcordata subcordata*
    (Herrich-Schäffer, 1869)

1121. *E. sempa* Evans, 1955

1122. *Onophas columbaria flossites* (Butler, 1874)

1123. *Styriodes badius* (Bell, 1930)

1124. *S. quaka* Evans, 1955

1125. *Styriodes* sp. n.

1126. *Enosis pruinosa pruinosa* (Plötz, 1882)

1127. *E. iccius* Evans, 1955

1128. *E. blotta* Evans, 1955

1129. *E. immaculata demon* Evans, 1955

1130. *E. angularis angularis* (Möschler, 1877)

1131. *Enosis* (?) sp.

1132. *Vertica verticalis* ssp. n.

1133. *Ebusus ebusus ebusus* (Cramer, 1780)

1134. *Talides sinois sinois* Hübner, 1819

1135. *Tromba tromba* Evans, 1955

1136. *Carystus jolus* (Cramer, 1782)

1137. *C. senex* (Plötz, 1882)

1138. *C. phorcus phorcus* (Cramer, 1777)

1139. *Tisias quadrata quadrata*
    (Herrich-Schäffer, 1869)

1140. *T. lesueur canna* Evans, 1955

1141. *Moeros moeros* (Möschler, 1877)

1142. *Cobalus virbius virbius* (Cramer, 1777)

1143. *C. calvina* (Hewitson, 1866)

1144. *Dubiella fiscella fiscella* (Hewitson, 1877)

1145. *D. dubius* (Cramer, 1781)

1146. *Tellona variegata* (Hewitson, 1870)

1147. *Damas clavus* (Herrich-Schäffer, 1869)

1148. *Orphe vatinius* Godman, 1901

1149. *O. gerasa* (Hewitson, 1867)

1150. *Carystoides basoches* (Latreille, 1824)

1151. *C. yenna* Evans, 1955

1152. *C. noseda* (Hewitson, 1866)

1153. *C. sicania orbius* (Godman, 1901)

1154. *C. cathaea* (Hewitson, 1866)

1155. *Perichares philetes philetes* (Gmelin, 1790)

1156. *P. lotus* (Butler, 1870)

1157. *Orses cynisca* (Swainson, 1821)

1158. *Alera* sp. n.

1159. *Lycas godart boisduvalii* (Ehrmann, 1909)

1160. *L. argentea* (Hewitson, 1866)

1161. *Saturnus saturnus saturnus* (Fabricius, 1787)

1162. *S. metonidia* (Schaus, 1902)

1163. *S. reticulata meton* (Mabille, 1891)

1164. *Phlebodes pertinax* (Cramer, 1781)

1165. *P. campo sifax* Evans, 1955

1166. *P. notex* Evans, 1955

1167. *P. vira* (Butler, 1870)

1168. *P. virgo* Evans, 1955 (?)

1169. *P. torax* Evans, 1955

1170. *Phlebodes* sp. n. 1

1171. *Phlebodes* sp. n. 2

1172. *Joanna boxi* Evans, 1955

1173. *Quinta cannae* (Herrich-Schäffer, 1869)

1174. *Cynea iquita* (Bell, 1941)

1175. *C. corisana* (Möschler, 1883)

1176. *C. popla* Evans, 1955

1177. *C. bistrigula* (Herrich-Schäffer, 1869)

1178. *C. diluta* (Herrich-Schäffer, 1869)

1179. *Penicula bryanti* (Weeks, 1906)

1180. *P. advena advena* (Draudt, 1923)

1181. *P. crista* Evans, 1955

1182. *Decinea decinea derisor* (Mabille, 1891)

1183. *Decinea* sp. n.

1184. *Orthos orthos orthos* (Godman, 1900)

1185. *O. minka* Evans, 1955

1186. *O. trinka* Evans, 1955

1187. *O. potesta* (Bell, 1941)

1188. *Conga chydaea* (Butler, 1877)

1189. *Hylephila phyleus phyleus* (Drury, 1773)

1190. *Wallengrenia curassavica* (Snellen, 1887)

1191. *Pompeius pompeius* (Latreille, 1824)

1192. *Mellana* sp. n.

1193. *M. meridiani pandora* (Hayward, 1940)

1194. *M. barbara* (Williams & Bell, 1931)

1195. *Arotis pandora* (Lindsey, 1925)

1196. *Hansa devergens devergens* (Draudt, 1923)

1197. *Metron leucogaster ambrosei* (Weeks, 1906)

1198. *M. zimra* (Hewitson, 1877)

1199. *Propertius propertius* (Fabricius, 1793)

1200. *Phemiades pohli pohli* (Bell, 1932)

1201. *Calpodes ethlius* (Cramer, 1782)

1202. *Panoquina ocola* (Edwards, 1863)

1203. *P. fusina fusina* (Hewitson, 1868)

1204. *Zenis jebus* ssp. n.

1205. *Lindra* sp. n.

1206. *L. boliviana* Mielke, 1993

1207. *Oxynthes corusca* (Herrich-Schäffer, 1869)

1208. *Niconiades xanthaphes* Hübner, 1821

1209. *N. linga* Evans, 1955

1210. *Aides duma argyrina* Cowan, 1970

1211. *A. brino* (Cramer, 1781)

1212. *A. aegita* (Hewitson, 1866)

1213. *Saliana triangularis* (Kaye, 1914)

1214. *S. fusta* Evans, 1955

1215. *S. fischer* (Latreille, 1824)

1216. *S. esperi* Evans, 1955

1217. *S. antoninus* (Latreille, 1824)

1218. *S. longirostris* (Sepp, 1840)

1219. *S. morsa* Evans, 1955

1220. *S. salius* (Cramer, 1775)

1221. *S. saladin culta* Evans, 1955

1222. *Thracides cleanthes telmela* (Hewitson, 1866)

1223. *T. thrasea* (Hewitson, 1866)

1224. *T. nanea nanea* (Hewitson, 1867)

1225. *T. phidon* (Cramer, 1779)

1226. *Neoxeniades cincia* (Hewitson, 1867)

1227. *N. braesia braesia* (Hewitson, 1867)

1228. *Neoxeniades* sp. n.
   (nr. *cincia* Hewitson, 1867)

1229. *Aroma aroma* (Hewitson, 1867)

1230. *Chloeria psittacina*
   (C & R Felder, 1867) (SR)

1231. *Pyrrhopygopsis socrates orasus*
   (Druce, 1876)

1232. Unidentified 1 (*Vehilius* ?)

1233. Unidentified 2 (nr. *Orphe*)

1234. Unidentified 3 (nr. *Psoralis*)

(SR) Sight record only

List of Butterflies From Pampas del Heath

Gerardo Lamas

NYMPHALIDAE		
Heliconiinae		
Heliconiini		
*	1. *Philaethria pygmalion* (Fruhstorfer, 1912)	GF
	2. *Dryadula phaetusa* (Linnaeus, 1758)	B (SR)
	3. *Dryas iulia alcionea* (Cramer, 1779)	F, GF
	4. *Neruda aoede manu* (Lamas, 1976)	GF
	5. *Heliconius burneyi ada* Neustetter, 1925	F
	6. *H. erato luscombei* Lamas, 1976	F, GF
	7. *H. melpomene schunkei* Lamas, 1976	F (SR)
	8. *H. numata lyrcaeus* Weymer, 1891	F, GF, PP
	9. *H. sara thamar* (Hübner, 1806)	F, GF
	10. *H. wallacei flavescens* Weymer, 1891	F
Nymphalinae		
Kallimini		
	11. *Junonia genoveva occidentalis* C. & R. Felder, 1862	GF, PP
Limenitidinae		
Coeini		
	12. *Historis odius* ssp. n.	F
	13. *Baeotus deucalion* (C. & R. Felder, 1860)	B (SR)
	14. *Colobura dirce dirce* (Linnaeus, 1758)	F
	15. *Tigridia acesta tapajona* (Butler, 1873)	F, GF (SR)
Biblidini		
	16. *Catonephele acontius acontius* (Linnaeus, 1771)	GF
	17. *C. numilia numilia* (Cramer, 1775)	GF
	18. *Nessaea obrina lesoudieri* LeMoult, 1933	F, GF
	19. *Eunica alpais alpais* (Godart, 1824)	GF
	20. *E. amelia erroneata* Oberthür, 1916	GF
	21. *E. malvina malvina* Bates, 1864	F
	22. *E. pusilla* Bates, 1864	GF
	23. *E. sydonia sydonia* (Godart, 1824)	F
	24. *Ectima iona* Doubleday, 1848	F
	25. *Pyrrhogyra crameri hagnodorus* Fruhstorfer, 1908	F
	26. *Temenis laothoe laothoe* (Cramer, 1777)	GF
	27. *Dynamine athemon barreiroi* Fernández, 1928	F

	28. *D. myrson* (Doubleday, 1849)	F
	29. *Diaethria clymena peruviana* (Guenée, 1872)	F (SR)
	30. *Callicore cynosura cynosura* (Doubleday, 1847)	F
*	31. *C. sorana horstii* (Mengel, 1916)	GF, PP

*	New record for Perú
B	Riverbanks
F	Forest
GF	Gallery Forest
PP	Pampas
(SR)	Sight record only

Limenitidini

	32. *Adelpha aethalia davisii* (Butler, 1867)	F
	33. *A. cocala urraca* (C. & R. Felder, 1862)	GF
	34. *A. cytherea lanilla* Fruhstorfer, 1913	GF
	35. *A. iphiclus iphiclus* (Linnaeus, 1758)	F, GF
	36. *A. jordani* Fruhstorfer, 1913	F
	37. *A. mesentina chancha* Staudinger, 1886	GF

Cyrestidini

	38. *Marpesia chiron marius* (Cramer, 1779)	GF, PP
	39. *M. crethon* (Fabricius, 1776)	B (SR)
	40. *M. egina* (Bates, 1865)	F, GF

Charaxinae

	41. *Polygrapha xenocrates xenocrates* (Westwood, 1850)	GF, PP
	42. *Memphis moruus memphis* (C. & R. Felder, 1867)	GF
	43. *Archaeoprepona demophon muson* (Fruhstorfer, 1905)	F
	44. *Prepona laertes demodice* (Godart, 1824)	GF

Apaturinae

	45. *Doxocopa agathina agathina* (Cramer, 1777)	F, GF
	46. *D. pavon pavon* (Latreille, 1809)	F (SR)

Morphinae

	47. *Antirrhea hela* C. & R. Felder, 1862	F
	48. *A. philoctetes avernus* Hopffer, 1874	F (SR)
	49. *Morpho achilles theodorus* Fruhstorfer, 1907	F, GF
	50. *M. deidamia grambergi* Weber, 1944	GF
	51. *M. hecuba cisseistricta* LeMoult & Réal, 1962	GF

Brassolinae

	52. *Opsiphanes cassiae strophios* Fruhstorfer, 1907	F
	53. *Opoptera aorsa hilara* Stichel, 1902	F (SR)
	54. *Catoblepia berecynthia adjecta* Stichel, 1906	F (SR)
	55. *C. soranus* Westwood, 1851	F
	56. *Caligo teucer semicaerulea* Joicey & Kaye, 1917	F

Satyrinae

	57. *Haetera piera* ssp. n.	F (SR)
	58. *P. astyoche boliviana* Brown, 1948	F
	59. *P. hortona albofasciata* Rosenberg & Talbot, 1914	F
	60. *P. hyalina extincta* Weymer, 1910	F
	61. *P. lamia chalybaea* Godman, 1905	F
	62. *P. lena brasiliensis* (C. & R. Felder, 1862)	F (SR)
	63. *Bia actorion rebeli* Bryk, 1953	F (SR)
	64. *Taygetis echo koepckei* Forster, 1964	F
	65. *T. sosis* Hopffer, 1874	GF
	66. *T. thamyra* (Cramer, 1779)	GF
	67. *Caeruleuptychia aegrota* (Butler, 1867)	PP
	68. *Chloreuptychia arnaca* (Fabricius, 1776)	F
	69. *C. catharina* (Staudinger, 1886)	F
	70. *C. herseis* (Godart, 1824)	GF
	71. *Euptychia* nr. *jesia* Butler, 1869	GF
	72. *Hermeuptychia hermes* (Fabricius, 1775)	GF, PP
*	73. *Hermeuptychia* sp. (n.?)	GF, PP
	74. *Magneuptychia gera nobilis* (Weymer, 1911)	F
	75. *M. ocypete* (Fabricius, 1776)	GF
	76. *Megeuptychia antonoe* (Cramer, 1775)	GF
	77. *Pareuptychia ocirrhoe* (Fabricius, 1776)	F
*	78. *Paryphthimoides melobosis* (Capronnier, 1874)	GF, PP
*	79. *Praefaunula* sp. n.	PP
*	80. *Yphthimoides electra* (Butler, 1867)	PP

Danainae

	81. *Lycorea halia pales* C. & R. Felder, 1862	GF

Ithomiinae

	82. *Melinaea marsaeus clara* Rosenberg & Talbot, 1914	GF
	83. *Thyridia psidii ino* C. & R. Felder, 1862	GF
	84. *Sais rosalia badia* Haensch, 1905	F, GF
	85. *Forbestra olivencia aeneola* Fox, 1967	F
	86. *Methona confusa psamathe* Godman & Salvin, 1898	GF
	87. *Napeogenes pharo pharo* (C. & R. Felder, 1862)	F
	88. *Oleria onega lentita* Lamas, 1985	F

	89. *Ithomia lichyi neivai* d'Almeida, 1940	F
	90. *Hypoleria virginia vitiosa* Lamas, 1985	F
	91. *Heterosais nephele nephele* (Bates, 1862)	F

RIODINIDAE

	92. *Euselasia mirania* (Bates, 1868)	F
	93. *E. teleclus* (Stoll, 1787)	F, GF
	94. *Semomesia tenella tenella* Stichel, 1910	F
	95. *Mesosemia cyanira* Stichel, 1909	GF
	96. *M. ibycus* Hewitson, 1859	GF
	97. *M. naiadella naiadella* Stichel, 1909	F, GF
	98. *M. nerine* Stichel, 1909	GF
	99. *M. sirenia sirenia* Stichel, 1909	F, GF
	100. *M. judicialis* Butler, 1874	F
	101. *Eurybia caerulescens* ssp.	F (SR)
*	102. *Alesa prema* (Godart, 1824)	F
	103. *A. hemiurga* Bates, 1867	F
	104. *Cremna actoris meleagris* Hopffer, 1874	GF
	105. *Ancyluris meliboeus meliboeus* (Fabricius, 1777)	F
	106. *Metacharis regalis regalis* Butler, 1867	F
*	107. *Cariomothis* sp. n.	F
	108. *Amarynthis meneria* (Cramer, 1776)	F, GF (SR)
*	109. *Lemonias glaphyra* (?*modesta*) (Mengel, 1902)	PP
*	110. *Audre middletoni* (Sharpe, 1890)	PP
	111. *Sarota acantus* (Cramer, 1781)	GF
	112. *Adelotypa mollis asemna* (Stichel, 1910)	GF
	113. *Setabis flammula* (Bates, 1868)	GF
	114. *Synargis gela gela* (Hewitson, 1853)	F
	115. *Nymphidium mantus* (Cramer, 1775)	GF
	116. *N. minuta* Druce, 1904	F
	117. *N. leucosia medusa* Druce, 1904	F
	118. *N. caricae parthenium* Stichel, 1924	F

LYCAENIDAE

	119. *Theritas mavors* Hübner, 1818	F
*	120. *Strymon cyanofusca* Johnson, Eisele & McPherson, 1990	PP
*	121. *S. tegaea* (Hewitson, 1868)	PP

*	New record for Perú
B	Riverbanks
F	Forest
GF	Gallery Forest
PP	Pampas
(SR)	Sight record only

	122. *S. megarus* (Godart, 1824)	PP
*	123. *S. serapio* (Godman & Salvin, 1887)	GF
*	124. *Lamprospilus badaca* (Hewitson, 1868)	PP
	125. *Calycopis atnius* (Herrich-Schäffer, 1853) complex	PP

PIERIDAE

Coliadinae

	126. *Anteos menippe* (Hübner, 1818)	F, PP (SR)
	127. *Aphrissa* sp.	B (SR)
	128. *Phoebis argante larra* (Fabricius, 1798)	B (SR)
	129. *P. philea philea* (Linnaeus, 1763)	B (SR)
	130. *P. sennae marcellina* (Cramer, 1777)	B (SR)
	131. *Rhabdodryas trite trite* (Linnaeus, 1758)	F (SR)
	132. *Glutophrissa drusilla drusilla* (Cramer, 1777)	F, PP
	133. *Itaballia demophile lucania* (Fruhstorfer, 1907)	F
	134. *Perrhybris pamela mazuka* Lamas, 1981	B (SR)

PAPILIONIDAE

	135. *Protographium agesilaus autosilaus* (Bates, 1861)	B (SR)
	136. *Heraclides thoas cinyras* (Ménétriès, 1857)	B (SR)

HESPERIIDAE

Pyrginae

	137. *Phanus vitreus* (Cramer, 1781)	F (SR)
	138. *Udranomia kikkawai* (Weeks, 1906)	GF
*	139. *U. spitzi* (Hayward, 1942)	GF, PP
	140. *Hyalothyrus leucomelas* (Geyer, 1832)	F
	141. *Phareas coeleste* Westwood, 1852	F (SR)
	142. *Epargyreus socus sinus* Evans, 1952	GF
	143. *Chrysoplectrum perniciosus perniciosus* (Herrich-Schäffer, 1869)	GF
	144. *Urbanus cindra* Evans, 1952	GF, PP
	145. *Oileides azines* (Hewitson, 1867)	F
	146. *Pachyneuria duidae duidae* (Bell, 1932)	GF
*	147. *Viola violella* (Mabille, 1898)	PP
	148. *Gorgythion begga pyralina* (Möschler, 1877)	GF
	149. *Zera tetrastigma erisichthon* (Plötz, 1884)	GF
	150. *Quadrus cerialis* (Cramer, 1782)	GF
*	151. *Q. fanda* Evans, 1953	GF

	152. *Q. deyrollei porta* Evans, 1953	F
	153. *Gindanes brebisson phagesia* (Hewitson, 1868)	GF
	154. *Pythonides jovianus fabricii* Kirby, 1871	GF
	155. *P. lerina* (Hewitson, 1868)	F
	156. *Milanion hemes* ssp.	GF
*	157. *Cogia hassan evansi* Bell, 1937	PP
	158. *Achyodes mithridates thraso* (Hübner, 1807)	F
	159. *Anastrus sempiternus simplicior* (Möschler, 1877)	F, GF
	160. *A. tolimus robigus* (Plötz, 1884)	F, GF
	161. *Camptopleura auxo* (Möschler, 1879)	GF
	162. *Chiomara basigutta* (Plötz, 1884)	PP
	163. *Heliopetes arsalte arsalte* (Linnaeus, 1758)	PP

Hesperiinae

	164. *Corticea corticea* (Plötz, 1882)	GF, PP
*	165. *Corticea* sp. (n.?)	PP
	166. *Artines aepitus* (Geyer, 1832)	F
*	167. *Artines* sp. n.	GF
*	168. *Vidius nappa* Evans, 1955	GF, PP
*	169. *V. anna* (Mabille, 1898)	GF
	170. *Cymaenes hazarma* (Hewitson, 1877)	GF
	171. *Mnasitheus chrysophrys* (Mabille, 1891)	F
	172. *Parphorus storax storax* (Mabille, 1891)	F
	173. *P. decora* (Herrich-Schäffer, 1869)	F
	174. *Papias phainis* Godman, 1900	GF
	175. *Cobalopsis nero* (Herrich-Schäffer, 1869)	F
	176. *Vettius richardi* (Weeks, 1906)	F
*	177. *V. lucretius* (Latreille, 1824)	GF
	178. *Turesis basta* Evans, 1955	F
	179. *Justinia justinianus dappa* Évans, 1955	F
	180. *Ebusus ebusus ebusus* (Cramer, 1780)	GF
	181. *Carystus hocus* Evans, 1955	GF
	182. *Carystoides yenna* Evans, 1955	F
	183. *C. cathaea* (Hewitson, 1866)	F
*	184. *Decinea lucifer* (Hübner, 1831)	GF
*	185. *Copaeodes castanea* Mielke, 1969	PP

*	New record for Perú
B	Riverbanks
F	Forest
GF	Gallery Forest
PP	Pampas
(SR)	Sight record only

	186. *Pompeius dares* (Plötz, 1883)	PP
	187. *P. amblyspila* (Mabille, 1898)	PP
*	188. *Mellana* sp.	PP
	189. *Perichares philetes philetes* (Gmelin, 1791)	F
*	190. *Euphyes cherra* Evans, 1955	PP
	191. *Arotis derasa derasa* (Herrich-Schäffer, 1870)	PP
*	192. *Chalcone* sp. n.	GF
	193. *Panoquina hecebolus* (Scudder, 1872)	PP
	194. *P. trix* Evans, 1955	GF
*	195. *P. bola* Bell, 1942	GF
	196. *P. fusina fusina* (Hewitson, 1868)	GF
	197. *Niconiades xanthaphes* Hübner, 1821	GF
	198. *Aides duma argyrina* Cowan, 1970	F
	199. *A. aegita* (Hewitson, 1866)	GF
	200. *Saliana chiomara* (Hewitson, 1867)	F
	201. *S. longirostris* (Sepp, 1840)	F
	202. *Thracides cleanthes trebla* Evans, 1955	GF
	203. *Pyrrhopygopsis socrates orasus* (Druce, 1876)	GF

CONSERVATION INTERNATIONAL